Magnetic Sensors and Devices
Technologies and Applications

Devices, Circuits, and Systems

Series Editor
Krzysztof Iniewski
Emerging Technologies CMOS Inc.
Vancouver, British Columbia, Canada

PUBLISHED TITLES:

Advances in Imaging and Sensing
Shuo Tang and Daryoosh Saeedkia

Analog Electronics for Radiation Detection
Renato Turchetta

Atomic Nanoscale Technology in the Nuclear Industry
Taeho Woo

Biological and Medical Sensor Technologies
Krzysztof Iniewski

Building Sensor Networks: From Design to Applications
Ioanis Nikolaidis and Krzysztof Iniewski

Cell and Material Interface: Advances in Tissue Engineering, Biosensor, Implant, and Imaging Technologies
Nihal Engin Vrana

Circuits and Systems for Security and Privacy
Farhana Sheikh and Leonel Sousa

Circuits at the Nanoscale: Communications, Imaging, and Sensing
Krzysztof Iniewski

CMOS: Front-End Electronics for Radiation Sensors
Angelo Rivetti

CMOS Time-Mode Circuits and Systems: Fundamentals and Applications
Fei Yuan

Design of 3D Integrated Circuits and Systems
Rohit Sharma

Diagnostic Devices with Microfluidics
Francesco Piraino and Šeila Selimović

Electrical Solitons: Theory, Design, and Applications
David Ricketts and Donhee Ham

Electronics for Radiation Detection
Krzysztof Iniewski

PUBLISHED TITLES:

Electrostatic Discharge Protection: Advances and Applications
Juin J. Liou

**Embedded and Networking Systems:
Design, Software, and Implementation**
Gul N. Khan and Krzysztof Iniewski

Energy Harvesting with Functional Materials and Microsystems
Madhu Bhaskaran, Sharath Sriram, and Krzysztof Iniewski

Gallium Nitride (GaN): Physics, Devices, and Technology
Farid Medjdoub

**Graphene, Carbon Nanotubes, and Nanostuctures:
Techniques and Applications**
James E. Morris and Krzysztof Iniewski

High-Speed Devices and Circuits with THz Applications
Jung Han Choi

High-Speed Photonics Interconnects
Lukas Chrostowski and Krzysztof Iniewski

**High Frequency Communication and Sensing:
Traveling-Wave Techniques**
Ahmet Tekin and Ahmed Emira

**High Performance CMOS Range Imaging:
Device Technology and Systems Considerations**
Andreas Süss

Integrated Microsystems: Electronics, Photonics, and Biotechnology
Krzysztof Iniewski

Integrated Power Devices and TCAD Simulation
Yue Fu, Zhanming Li, Wai Tung Ng, and Johnny K.O. Sin

Internet Networks: Wired, Wireless, and Optical Technologies
Krzysztof Iniewski

Introduction to Smart eHealth and eCare Technologies
Sari Merilampi, Krzysztof Iniewski, and Andrew Sirkka

Ionizing Radiation Effects in Electronics: From Memories to Imagers
Marta Bagatin and Simone Gerardin

Labs on Chip: Principles, Design, and Technology
Eugenio Iannone

Laser-Based Optical Detection of Explosives
Paul M. Pellegrino, Ellen L. Holthoff, and Mikella E. Farrell

Low Power Emerging Wireless Technologies
Reza Mahmoudi and Krzysztof Iniewski

PUBLISHED TITLES:

Magnetic Sensors and Devices: Technologies and Applications
Kirill Poletkin and Laurent A. Francis

Medical Imaging: Technology and Applications
Troy Farncombe and Krzysztof Iniewski

Metallic Spintronic Devices
Xiaobin Wang

MEMS: Fundamental Technology and Applications
Vikas Choudhary and Krzysztof Iniewski

Micro- and Nanoelectronics: Emerging Device Challenges and Solutions
Tomasz Brozek

Microfluidics and Nanotechnology: Biosensing to the Single Molecule Limit
Eric Lagally

MIMO Power Line Communications: Narrow and Broadband Standards, EMC, and Advanced Processing
Lars Torsten Berger, Andreas Schwager, Pascal Pagani, and Daniel Schneider

Mixed-Signal Circuits
Thomas Noulis

Mobile Point-of-Care Monitors and Diagnostic Device Design
Walter Karlen

Multisensor Attitude Estimation: Fundamental Concepts and Applications
Hassen Fourati and Djamel Eddine Chouaib Belkhiat

Multisensor Data Fusion: From Algorithm and Architecture Design to Applications
Hassen Fourati

MRI: Physics, Image Reconstruction, and Analysis
Angshul Majumdar and Rabab Ward

Nano-Semiconductors: Devices and Technology
Krzysztof Iniewski

Nanoelectronic Device Applications Handbook
James E. Morris and Krzysztof Iniewski

Nanomaterials: A Guide to Fabrication and Applications
Sivashankar Krishnamoorthy

Nanopatterning and Nanoscale Devices for Biological Applications
Šeila Selimović

Nanoplasmonics: Advanced Device Applications
James W. M. Chon and Krzysztof Iniewski

Nanoscale Semiconductor Memories: Technology and Applications
Santosh K. Kurinec and Krzysztof Iniewski

PUBLISHED TITLES:

Noise Coupling in System-on-Chip
Thomas Noulis

Novel Advances in Microsystems Technologies and Their Applications
Laurent A. Francis and Krzysztof Iniewski

Optical, Acoustic, Magnetic, and Mechanical Sensor Technologies
Krzysztof Iniewski

Optical Fiber Sensors: Advanced Techniques and Applications
Ginu Rajan

Optical Imaging Devices: New Technologies and Applications
Ajit Khosla and Dongsoo Kim

Organic Solar Cells: Materials, Devices, Interfaces, and Modeling
Qiquan Qiao

Physical Design for 3D Integrated Circuits
Aida Todri-Sanial and Chuan Seng Tan

Power Management Integrated Circuits and Technologies
Mona M. Hella and Patrick Mercier

Radiation Detectors for Medical Imaging
Jan S. Iwanczyk

Radiation Effects in Semiconductors
Krzysztof Iniewski

Reconfigurable Logic: Architecture, Tools, and Applications
Pierre-Emmanuel Gaillardon

Semiconductor Devices in Harsh Conditions
Kirsten Weide-Zaage and Malgorzata Chrzanowska-Jeske

Semiconductor Radiation Detection Systems
Krzysztof Iniewski

Semiconductor Radiation Detectors, Technology, and Applications
Salim Reza

Semiconductors: Integrated Circuit Design for Manufacturability
Artur Balasinski

Smart Grids: Clouds, Communications, Open Source, and Automation
David Bakken

Smart Sensors for Industrial Applications
Krzysztof Iniewski

Soft Errors: From Particles to Circuits
Jean-Luc Autran and Daniela Munteanu

Solid-State Radiation Detectors: Technology and Applications
Salah Awadalla

PUBLISHED TITLES:

Structural Health Monitoring of Composite Structures Using Fiber Optic Methods
Ginu Rajan and Gangadhara Prusty

Technologies for Smart Sensors and Sensor Fusion
Kevin Yallup and Krzysztof Iniewski

Telecommunication Networks
Eugenio Iannone

Testing for Small-Delay Defects in Nanoscale CMOS Integrated Circuits
Sandeep K. Goel and Krishnendu Chakrabarty

Tunable RF Components and Circuits: Applications in Mobile Handsets
Jeffrey L. Hilbert

VLSI: Circuits for Emerging Applications
Tomasz Wojcicki

Wireless Medical Systems and Algorithms: Design and Applications
Pietro Salvo and Miguel Hernandez-Silveira

Wireless Technologies: Circuits, Systems, and Devices
Krzysztof Iniewski

Wireless Transceiver Circuits: System Perspectives and Design Aspects
Woogeun Rhee

FORTHCOMING TITLES:

3D Integration in VLSI Circuits: Design, Architecture, and Implementation Technologies
Katsuyuki Sakuma

Energy Efficient Computing: Devices, Circuits, and Systems
Santosh K. Kurinec

High-Speed and Lower Power Technologies: Electronics and Photonics
Jung Han Choi and Krzysztof Iniewski

IoT and Low-Power Wireless: Circuits, Architectures, and Techniques
Christopher Siu and Krzysztof Iniewski

Nanoelectronics: Devices, Circuits, and Systems
Nikos Konofaos

Radio Frequency Integrated Circuit Design
Sebastian Magierowski

Sensors for Diagnostics and Monitoring
Kevin Yallup and Krzysztof Iniewski

X-Ray Diffraction Imaging: Technology and Applications
Joel Greenberg and Krzysztof Iniewski

Magnetic Sensors and Devices
Technologies and Applications

Edited by
Kirill Poletkin and Laurent A. Francis

Managing Editor
Krzysztof Iniewski

CRC Press is an imprint of the
Taylor & Francis Group, an **informa** business

MATLAB® is a trademark of The MathWorks, Inc. and is used with permission. The MathWorks does not warrant the accuracy of the text or exercises in this book. This book's use or discussion of MATLAB® software or related products does not constitute endorsement or sponsorship by The MathWorks of a particular pedagogical approach or particular use of the MATLAB® software.

CRC Press
Taylor & Francis Group
6000 Broken Sound Parkway NW, Suite 300
Boca Raton, FL 33487-2742

© 2018 by Taylor & Francis Group, LLC

CRC Press is an imprint of Taylor & Francis Group, an Informa business

No claim to original U.S. Government works

Printed on acid-free paper

International Standard Book Number-13: 978-1-4987-1097-8 (Hardback)

This book contains information obtained from authentic and highly regarded sources. Reasonable efforts have been made to publish reliable data and information, but the author and publisher cannot assume responsibility for the validity of all materials or the consequences of their use. The authors and publishers have attempted to trace the copyright holders of all material reproduced in this publication and apologize to copyright holders if permission to publish in this form has not been obtained. If any copyright material has not been acknowledged, please write and let us know so we may rectify in any future reprint.

Except as permitted under U.S. Copyright Law, no part of this book may be reprinted, reproduced, transmitted, or utilized in any form by any electronic, mechanical, or other means, now known or hereafter invented, including photocopying, microfilming, and recording, or in any information storage or retrieval system, without written permission from the publishers.

For permission to photocopy or use material electronically from this work, please access www.copyright.com (http://www.copyright.com/) or contact the Copyright Clearance Center, Inc. (CCC), 222 Rosewood Drive, Danvers, MA 01923, 978-750-8400. CCC is a not-for-profit organization that provides licenses and registration for a variety of users. For organizations that have been granted a photocopy license by the CCC, a separate system of payment has been arranged.

Trademark Notice: Product or corporate names may be trademarks or registered trademarks, and are used only for identification and explanation without intent to infringe.

Visit the Taylor & Francis Web site at
http://www.taylorandfrancis.com

and the CRC Press Web site at
http://www.crcpress.com

Printed and bound in the United States of America by Sheridan

Contents

Series Editor xi
Editors xiii
Contributors xv

1 Biosensor Application for Bovine Mastitis Diagnosis 1
  Carla M. Duarte, Susana Cardoso, and Paulo P. Freitas

2 Giant (GMR) and Tunnel (TMR) Magnetoresistance Sensors:
  From Phenomena to Applications 35
  Càndid Reig and María-Dolores Cubells-Beltrán

3 Frequency Tuning Investigation of an Out-of-Plane Resonant
  Microstructure for a Capacitive Detection Magnetometer 65
  *Petros Gkotsis, Mohamed Hadj Said, Farès Tounsi, Brahim Mezghani,
  and Laurent A. Francis*

4 Micromachined Inductive Contactless Suspension: Technology
  and Modeling 101
  *Kirill V. Poletkin, Vlad Badilita, Zhiqiu Lu, Ulrike Wallrabe, and
  Christopher Shearwood*

5 Application of Magnetic Sensors for Ecological Monitoring
  of Stationary Ferromagnetic Masses from On Board Mobile
  Platforms 133
  *Alexander I. Chernomorsky, Vyacheslav E. Plekhanov, and
  Vladimir N. Maximov*

6 A Model to Calculate Force Characteristics of a Magnetic
  Suspension of a Superconducting Sphere 165
  Sergey I. Kuznetsov and Yury M. Urman

7 Magnetic Angle Sensors 201
  Udo Ausserlechner

Index 251

Series Editor

Krzysztof (Kris) Iniewski is managing R&D at Redlen Technologies Inc., a start-up company in Vancouver, Columbia, Canada. Redlen's revolutionary production process for advanced semiconductor materials enables a new generation of more accurate, all-digital, radiation-based imaging solutions. Kris is also a founder of Emerging Technologies CMOS Inc. (www.etcmos.com), an organization of high-tech events covering communications, microsystems, optoelectronics, and sensors. In his career, Dr. Iniewski has held numerous faculty and management positions at the University of Toronto, University of Alberta, SFU, and PMC-Sierra Inc. He has published over 100 research papers in international journals and conferences. He holds 18 international patents granted in the United States, Canada, France, Germany, and Japan. He is a frequent invited speaker and has consulted for multiple organizations internationally. He has written and edited several books for CRC Press, Cambridge University Press, IEEE Press, Wiley, McGraw-Hill, Artech House, and Springer. His personal goal is to contribute to healthy living and sustainability through innovative engineering solutions. In his leisurely time, Kris can be found hiking, sailing, skiing, or biking in beautiful British Columbia. He can be reached at kris.iniewski@gmail.com.

Editors

Kirill V. Poletkin is a research associate at the Institute of Microstructure Technology at Karlsruhe Institute of Technology. He received an honors diploma in electromechanical engineering, majoring in aviation devices and measurement systems, in 2001 from Nizhny Novgorod State Technical University, and his PhD from Moscow Aviation Institute (State University of Aerospace Technologies), Russia, in 2007. His master's thesis won the Ministry of Education and Science of the Russian Federation's best research work at an all-Russian student competition in 2002. Then, he was awarded the Humboldt Research Fellowship for Experienced Researchers in 2012 and successfully performed his Humboldt project titled "Micromachined contactless suspension with zero spring constant for application as an accelerometer" at University of Freiburg, Germany from 2013 until 2015. Between 2009 and 2013, he did postdoctoral research at Nanyang Technological University, Singapore. He has also worked previously with Giesecke & Devrient GmbH (G&D), JSC Temp-Avia, and the Russian Federal Nuclear Center (VNIITF). His research interests include micro- and nano-scales device and the processes of energy transfer within these scales. He is an author of a number of peer-reviewed journal articles, conference papers, and two chapters in books published by CRC Press.

Laurent A. Francis is a professor at the Institute of Information and Communication Technologies, Electronics and Applied Mathematics (ICTEAM) of the Université catholique de Louvain (UCL), Belgium. He received his degree in materials science engineering with a minor in electrical engineering in 2001, and in 2006 he received his PhD in applied science from UCL. His main focus is on the micro- and nanotechnology of co-integrated, ultra-low power CMOS MEMS sensors for biomedical applications and harsh environments. While working on his PhD, he worked as a researcher at IMEC in Leuven, Belgium, in the field of acoustic and optical biosensors and piezoelectric RF-MEMS. In 2007, he joined UCL as microsystems chair and created the Sensors, Microsystems and Actuators Laboratory of Louvain (SMALL), which he has led since. In 2011, he served as visiting professor at the Université de Sherbrooke, Canada, where he is currently professeur associé at the Département de génie électrique et de génie informatique. He has published more than 100 research papers in international journals, is co-editor of one book in microsystems technology published by CRC Press, and holds four patents. He served as TPC for conferences in the fields of harsh environments and materials science. He is a board member of the Belgian National Committee Biomedical Engineering and Belgian representative at IMEKO, the International Measurement Confederation. He is a regular member of the IEEE and serves as treasurer of the IEEE CPMT Benelux Chapter.

Contributors

Udo Ausserlechner
Infineon Technologies Austria AG
Villach, Austria

Vlad Badilita
Institute of Microstructure Technology
Karlsruhe Institute of Technology
Karlsruhe, Germany

Susana Cardoso
INESC—Microsistemas e Nanotecnologias
and
Instituto Superior Técnico, Physics Department
Universidade de Lisboa
Lisbon, Portugal

Alexander I. Chernomorsky
School of Control Systems, Informatics and Electrical Engineering
Moscow Aviation Institute
Moscow, Russia

María-Dolores Cubells-Beltrán
University of Valencia
Valencia, Spain

Carla M. Duarte
INESC—Microsistemas e Nanotecnologias
and
Faculdade de Medicina Veterinária
Lisbon, Portugal

Paulo P. Freitas
INESC—Microsistemas e Nanotecnologias
and
Instituto Superior Técnico, Physics Department
Universidade de Lisboa
Lisbon, Portugal

Petros Gkotsis
European Commission–Joint Research Centre
Seville, Spain

Sergey Kuznetsov
N. I. Lobachevsky State University of Nizhniy Novgorod
Nizhniy Novgorod, Russia
and
Johns Hopkins University
Baltimore, Maryland

Zhiqiu Lu
Department of Microsystems Engineering—IMTEK
University of Freiburg
Freiburg, Germany

Vladimir N. Maximov
School of Control Systems, Informatics and Electrical Engineering
Moscow Aviation Institute
Moscow, Russia

Brahim Mezghani
National Engineering School of Sfax
Sfax, Tunisia

Vyacheslav E. Plekhanov
School of Control Systems
 Informatics and Electrical
 Engineering
Moscow Aviation Institute
Moscow, Russia

Càndid Reig
University of Valencia
Valencia, Spain

Mohamed Hadj Said
National Engineering School of Sfax
Sfax, Tunisia

Christopher Shearwood
School of Mechanical & Aerospace
 Engineering
Nanyang Technological University
Singapore

Farès Tounsi
National Engineering School of Sfax
Sfax, Tunisia

Yury M. Urman
Nizhniy Novgorod Institute of
 Management and Business
Nizhniy Novgorod, Russia

Ulrike Wallrabe
Department of Microsystems
 Engineering—IMTEK
University of Freiburg
Freiburg, Germany

1
Biosensor Application for Bovine Mastitis Diagnosis

Carla M. Duarte, Susana Cardoso, and Paulo P. Freitas

CONTENTS

1.1 Introduction 2
    1.1.1 Bacterial Detection Methods 3
    1.1.2 Technological Advances 5
1.2 Portable "Lab-on-Chip" Platform for Bovine Mastitis Diagnosis in Raw Milk 5
    1.2.1 Magnetic Detection Principles 6
    1.2.2 Biosensor Fabrication 8
    1.2.3 Microfluidic Channel Fabrication 10
    1.2.4 Readout Electronics 10
    1.2.5 Biological Functionalization of Nanoparticles 11
    1.2.6 Biosensor Calibration 12
    1.2.7 Bacterial Cells 12
    1.2.8 Sterile Milk Samples 13
    1.2.9 Mastitic Milk Samples 13
    1.2.10 PCR Reference Method Analysis 13
    1.2.11 Biosensor Analysis 14
    1.2.12 Data Analysis 16
1.3 Results 17
    1.3.1 Evaluation of Biosensor's Bacterial Quantification 17
    1.3.2 Validation of Magnetic Detection with mAb Anti-*S. agalactiae* and pAb Anti-GB Streptococci 18
    1.3.3 Validation of Magnetic Detection with mAb Anti-*Staphylococcus* spp. and pAb Anti-*Staphylococcus aureus* 22
1.4 Discussion 24
    1.4.1 Strengths, Weaknesses, Opportunities, and Threats 27
1.5 Future Perspectives 30
References 30

ABSTRACT Bovine mastitis is an inflammation of the mammary gland, most often of infectious origin. It is the most frequent disease of dairy cattle and one of the main reasons for culling dairy cows (Gröhn et al., 1998; Hortet & Seegers, 1998; Hovi & Roderick, 1999). Bovine mastitis is also an economic burden for farmers because of decreased milk yield, cost of veterinary treatments, and other factors (Korhonen & Kaartinen, 1995). Dairy farm management focusing on animal health and hygiene improvement program implementation helps to control mastitis. The timely identification of causing microorganisms is necessary to control the disease, reduce the risk of chronic infections, and target the antimicrobial therapy to be used. Also, several studies showed that the early detection of mastitis may increase the cure rate by 60% and reduce the time required to recover normal milk production when combined with appropriate antimicrobial therapy (Milner et al., 1997).

1.1 Introduction

Bovine mastitis is an inflammation of the mammary gland, most often of infectious origin. It is the most frequent disease of dairy cattle and one of the main reasons for culling dairy cows (Gröhn et al., 1998; Hortet & Seegers, 1998; Hovi & Roderick, 1999). Bovine mastitis is also an economic burden on farmers because of decreased milk yields, cost of veterinary treatments, and other factors (Korhonen & Kaartinen, 1995). Dairy farm management focusing on animal health and hygiene improvement program implementation contributes to control mastitis. The timely identification of causative microorganisms is necessary to control the disease, reduce the risk of chronic infections, and target the antimicrobial therapy to be used. Also, several studies have shown that the early detection of mastitis may increase the cure rate by 60% and reduce the time required to recover normal milk production when combined with appropriate antimicrobial therapy (Milner et al., 1997). The rapid identification of pathogens such as *Staphylococcus* spp. and *Streptococcus* spp. and among these, the discrimination of major contagious pathogens *Staphylococcus aureus*, *Streptococcus agalactiae*, and *Streptococcus uberis* (Bradley, 2002; Zadoks et al., 2011), will therefore contribute to decreasing the economic burden of bovine mastitis. Coagulase-negative staphylococci, as *Staphylococcus epidermidis*, are considered minor mastitis pathogens, but they are the most common agents isolated from milk samples in several large-scale surveys worldwide (Tenhagen et al., 2006).

This work describes the application of a novel biosensor device for the identification and quantification of *S. agalactiae*, *S. uberis*, and *S. aureus* and of the genus *Staphylococcus* in raw milk samples. The biosensor consists of a magnetoresistive (MR) sensor and microfluidics, with electronic readout, and the detection of bacteria is achieved through the detection of magnetic nanoparticles (NP) labeling the bacteria.

1.1.1 Bacterial Detection Methods

Because of the high impact on human health, bacteria detection in milk has prompted many years of research and development toward reliable and universal diagnosis tests. Table 1.1 summarizes the different strategies and techniques used for the analysis of milk samples toward bacteria detection and quantification.

One of the most widely used methods for subclinical mastitis diagnosis is the California Mastitis Test (CMT), a common indirect method for somatic cell count (SCC) measurement. However, this portable method only discriminates sick from healthy animals and is unable to identify the causative agent of infection. Therefore, a microbiological culture is considered the gold standard for diagnosing mastitis pathogens (Britten, 2012), allowing for a targeted control and treatment decision, in addition to presenting high sensitivity and specificity. Another advantage of microbial culture–based methods is the possibility of identifying the antibiotic susceptibility of bacteria. The limitations of a microbiological culture include delays in obtaining results and suboptimal accuracy in identifying mastitis pathogens. On the other hand, the use of polymerase chain reaction (PCR)-based tests may be of interest for intramammary infection (IMI) diagnosis when milk samples with high SCC are culture-negative or when culturing only detects minor pathogens (Taponen et al., 2009; Bexiga et al., 2011). PCR is a semi-quantitative technique that generates information about the number of copies of DNA fragments that have been detected in a sample, being difficult to assign mastitis causality to a particular microorganism. We also cannot assume that one bacterial species is more important in terms of the negative effects on a mammary gland, simply because it is present in higher numbers.

Immunodiagnostics also create new perspectives for the diagnosis of bovine mastitis as an alternative to a microbiological culture. Methods based on serology have the desired characteristics for an ideal diagnostic test such as speed, sensitivity, ease of handling, and low cost (Fabres-Klein et al., 2014). The market already provides several portable commercialized immunoassays for the diagnosis of diseases of veterinary relevance (Zschöck et al., 2005).

The successful choice of a test that evaluates milk requires methodological knowledge and diagnostic capabilities for each test currently available. The sensitivity of culture tests may be complemented by PCR analysis, which are often combined to yield more robust results. However, to make treatment decisions, this combination does not allow for a timely answer. Proteomic research for reliable biomarkers, as enzymes and acute phase proteins (Pyörälä, 2003; Grönlund, et al., 2003; Åkerstedt et al., 2011; Mansor et al., 2013), is viable for the early detection of mastitis and drug efficacy, and to discover potentially novel targets for the development of alternative

TABLE 1.1
Overview of the Existing Strategies for Bacteria Detection in Milk

Method Basis for Bacteria Detection in Milk	LOD (Limit of Detection)	Sensitivity and Specificity	References
Microbiological			
On-farm culture test		Sensitivities of 97.9% and 93.8%	McCarron et al. (2009)
		Specificities of 68.6% and 70.1%	Lago et al. (2011)
Genetics			
Pathogen-specific targets of DNA were amplified and transferred to react and hybridize with specific probes that were pre-spotted on the biochip	10^3–10^5 CFU/mL	Sensitivity of 94.1%	Lee et al. (2008)
		Specificity of 100%	
Immunological			
Magnetic bead–based ELISA employing monoclonal antibodies for the detection of staphylococci in milk	10^4–10^5 CFU/mL		Yazdankhah et al. (1998)
Immunological			
ELISA for detecting *S. aureus* in milk samples	10^3–10^5 CFU/mL	Sensitivities of 69%–90%	Matsushita et al. (1990)
		Specificities of 61%–97%	Hicks et al. (1994)
Immunoagglutination			
S. aureus identification in milk samples		Sensitivity of 86.7%	Zschöck et al. (2005)
		Specificity of 90.1%	
Immunological			
Sandwich ELISA test to detect *S. aureus*	10^5 CFU/mL		Libing et al. (2012)
Immunological			
Competitive immunoassay performed by an amperometric magnetoimmunosensor for the specific detection and quantification of staphylococcal protein A and *S. aureus* cells	1 CFU/mL		de Ávila et al. (2012)

therapies (Lippolis & Reinhardt, 2010). However, these innovations are still not possible to use for routine diagnosis. Therefore, it remains important to develop a low-cost tool for the differentiation of clinically relevant mastitis pathogens that may be used on-farm.

1.1.2 Technological Advances

Flow cytometers have been optimized for use in portable platforms, where cell separation, identification, and counting can be achieved in a compact and modular format. In this work, this feature was combined with magnetic detection, where MR sensors were integrated within microfluidic channels to detect magnetically labeled cells.

Over the past years, the drawbacks of conventional flow cytometers have encouraged efforts to take advantage of microfabrication technologies and advanced microfluidics to achieve smaller, simpler, more innovative, and low-cost instrumentation with enhanced portability for on-site measurements. This miniaturization approach has in general made use of inexpensive polymers such as polydimethylsiloxane (PDMS) (Huh et al., 2005) and detection techniques easily integrated with electronics (Chung & Kim, 2007), such as MR sensors (Loureiro et al., 2011, Kokkinis et al., 2017). A previously reported work for magnetic particle detection (Loureiro et al., 2011; Freitas et al., 2012) used an integrated cytometer platform, consisting of MR sensors, readout/acquisition electronics, and a microfluidic channel where the sample with magnetic particles was injected.

Biosensors are fast becoming the next generation of tools in analyzing areas such as environmental research, medicine, biodefense, agriculture, and food control (Lazcka et al., 2007). The suitability of a detection method for routine diagnosis depends on its specificity, sensitivity, cost, processing time, and suitability for a large number of milk samples. New technical advances in mastitis diagnosis still require specialized training and experience to interpret results. The personnel responsible should be aware of the strict compliance to each step in the process for good quality control in obtaining reliable data.

1.2 Portable "Lab-on-Chip" Platform for Bovine Mastitis Diagnosis in Raw Milk

This work describes a platform for the in-flow detection of magnetically labeled cells with an MR-based cell cytometer. This portable device is composed of MR sensors (Freitas et al., 2007), namely, spin-valve sensors (SVs), integrated with a microfluidic platform and connected to an amplification and acquisition setup. The sensors have excellent spatial resolution (on the micrometer range) and are sensitive to the magnetic field created by magnetized beads flowing in PDMS microchannels above the sensors. The detection scheme used in this platform relies on the MR sensor's sensitivity to count individual cells in flow, contrary to other approaches (Mujika et al., 2009) while providing information on NPs' magnetization direction along the flow process. Therefore, no additional cell culture is needed. In addition, this platform is compatible with complex matrixes without the need for intricate sample preprocessing, while using a detection principle (magnetic) non-existent in nature (thus greatly reducing biological background noise and false positives). The use of MR sensors also simplifies the connection with electrical equipment while still allowing coupling with other detection techniques (e.g., fluorescence or a laser-irradiated magnetic bead system [LIMBS]) if needed. The sensors are sensitive to the magnetic field created by magnetically labeled cells flowing in microchannels above the sensors. This dynamic detection is based on immunoassay methodology since the selected specific antibodies (probes) recognize immunogenic proteins on bacteria cell walls (targets). The SV sensor detects the fringe field of the magnetic labels bound around the target cell through the specific probe.

1.2.1 Magnetic Detection Principles

SVs are MR sensors (Baibich et al., 1988; Dieny, 1994) consisting of a non-magnetic (NM) metal film between two layers of ferromagnetic (FM) metals. The film thicknesses are of nanometer dimensions (1–3 nm), and the sensor mechanism is based on the property of the magnetic layer to align with an external magnetic field. Since the resistance of these materials depends on the relative magnetic orientations of the FM layers (from parallel to anti-parallel), the SV is the optimum transductor between magnetic field and resistance. Figure 1.1 shows an example of an SV sensor, microfabricated into a 10×2 μm^2 active area, with a linear response as a function of the external

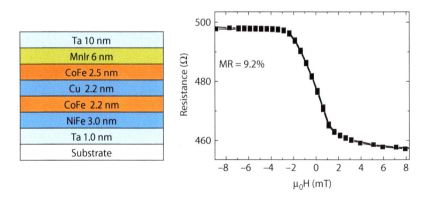

FIGURE 1.1
Typical structure for a magnetoresistive spin-valve stack and resistance vs. magnetic field transfer curve of a spin-valve magnetoresistive sensor.

magnetic field. Many successful examples of SV sensors have been published over the past years, from magnetic recording, memories, biomedical, automotive, among many other applications (Loureiro et al., 2011; Freitas et al., 2012; Wang & Li, 2008). Thus, SV sensors are a mature technology with a large community of industrial and academic users. As a consequence, large-scale integration is offered by manufacturers to biomedical and biotechnological researchers toward hybrid integration with microfluidic platforms, for example.

The dimensions of the SV sensors are optimized taking into account the final sensor application. In biological applications, the detection targets have sizes ranging between a few nanometers (nm; molecules such as DNA, RNA, and various proteins) to tens of micrometers (μm; cells can vary in size from 1 μm, like the target cell described in this chapter, to 100 μm, the size of a big plant eukaryotic cell). Detection is performed through the magnetic labeling of these biological targets with nano- or micrometer superparamagnetic particles that, under a magnetic field, acquire a magnetic moment. This creates a fringe field that can be detected by the sensor through its changed resistance.

The sensor output can be written as

$$\Delta V = -S \times R_{sq} \times I \times \left(H_{ext} + H_{bias} + H_{coup}\right) \quad (1.1)$$

where:
H_{bias} is the bias field used to center the SV transfer curve (e.g., Oersted field created by the electrical current)
H_{coup} includes the internal FM and magnetostatic fields across the thin film layers of the SV stack
S = MR/(2 $(H_k + H_{demag})$) is the sensor's sensitivity
R_{sq} = $R \times L/W$ is the sensor's resistance accounting for the dimensions (L = length, W = width)
H_{ext} represents the external magnetic field, averaged over the sensor's area

In our case, this is the fringe field created by the magnetic labels (Freitas et al., 2000, 2012, 2007). Normally, the NP used for biomedical applications are weakly magnetic or paramagnetic, to avoid clustering inside the microchannels. Therefore, the sensors are required to read small magnetic fields (few micro to nanotesla).

The dynamic detection approach employed in this work involves the application of a magnetic field perpendicular to the sensors in order to magnetize the beads labeling the cells, with minimum impact on the in-plane sensing direction of the sensor (as will be described in Section 1.2.2.). The dynamic detection mechanism is illustrated in Figure 1.2, where a vertically magnetized particle is injected through a microchannel and generates a variable field over the sensor. In Position 1, because of the large distance to the sensor, the fringe field produced by the particles is negligible. As the particle

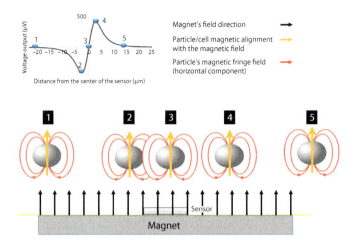

FIGURE 1.2
Schematics of MR sensor detection of magnetically labeled targets flowing above the sensor, from left (Position 1) to right (Position 5). (Adapted from Fernandes, A.C. et al., *Sensors*, 14, 15496–15524, 2014.)

approaches the sensor, the free layer will sense the right-side component of the particles' fringe field, which changes the sensor resistance (Position 2). When the particle is in the center of the sensor, the average fringe field of the particle is equal to zero, thereby vanishing the signal (Position 3). Finally, as the particle passes the sensor (Position 4), the free layer magnetization is affected by an opposite fringe field component when compared to Position 2. When the cells go away, the signal goes back to zero since no fringe field is sensed (Position 5). As a result, a bipolar peak is the signature of the passage of a perpendicularly magnetized particle over the SV sensor.

In a dynamic approach, sample acquisition velocity depends on the electronics, thus allowing a high throughput and the direct number of cells to the number of signals relation. A dynamic acquisition requires magnetic labels with a high magnetic moment under an applied external magnetic field to obtain a large detectable fringe field, significantly larger than the noise background level. However, there needs to be careful label selection, as these should possess a non-remnant moment in order to avoid particle clustering during the labeling process of the cells, which can result in cell clustering creating an underestimation of the cell count. A reduced label size is also important to avoid the detection of isolated particles (Freitas et al., 2012; Wang & Li, 2008).

1.2.2 Biosensor Fabrication

Magnetic detection in a cytometer biochip platform was first reported in 2007 (Loureiro et al., 2007). In this and subsequent work (Loureiro et al., 2011; Freitas et al., 2012), magnetic particle detection was done with an integrated cytometer platform, consisting of MR sensors and readout/acquisition electronics and a microfluidic channel where the magnetically labeled sample (in this case, the milk) was injected. The device geometry and physical principles of operation are described (Fernandes et al., 2014) and are based on SVs deposited by ion beam deposition on a Nordiko 3000 tool (Gehanno et al., 1999) with the following structure: Si/Al$_2$O$_3$ 60/Ta 1.5/Ni$_{80}$Fe$_{20}$ 2.5/Co$_{90}$Fe$_{10}$ 2.0/Cu 2.1/Co$_{90}$Fe$_{10}$ 2.0/Mn$_{76}$Ir$_{24}$ 6.0/Ta 5.0 (Freitas et al., 2007) (thickness in nm, compositions in atomic %), patterned with 3 × 100 µm active dimensions (measured between the AlSiCu 300 nm thick contact leads), according to Figure 1.3. Passivation was done with a 300 nm thick Si$_3$N$_4$ layer deposited by a plasma-enhanced chemical vapor deposition (PECVD; Electrotech Delta chemical vapor deposition system). Sensors were annealed at 250°C for 15 min, in vacuum, and cooled down under 1 T magnetic field.

The SV sensors' electrical transport characterization (resistance vs. direct current [dc] magnetic field) provided information on the magnetoresistance, defined as MR = $(R_{max} - R_{min})/R_{min}$ (where R_{max} and R_{min} are the maximum and minimum resistance levels). The sensor sensitivity is defined as the slope of the curve over the linear range of operation and ranges 0.15%–0.17%/Oe for the sensors measured across the wafer.

The magnetic detection mechanism used NP that had a superparamagnetic signature, thereby requiring an external magnetic field to activate their magnetization. This was done with an external vertical field created by a permanent magnet (NdFeB, 20-10-01STIC, Supermagnete) mounted below the printed circuit board (PCB). After magnet alignment below this sensor, the effect of the small components in the plane of the sensors was visible in their sensitivity decrease to 0.074%/Oe. The magnetic field at the microfluidic channel center was ~31 mT, so the individual NP were magnetized with a magnetic moment of 2.0×10^{-18} Am^2. Upon magnetization, the NP created a magnetic field fringe field at the sensor surface; therefore, the

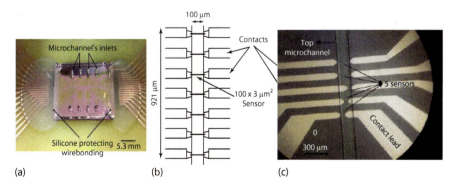

FIGURE 1.3
(a) Device with the magnetoresistive chip bonded to the polydimethylsiloxane (PDMS) microchannels. The sensor's wire bonding wires are protected with silicone. (b) Spin-valve (SV) sensor distribution along the microchannels. (c) Microscope photo of the fabricated SVs with the PDMS microchannel over them (20× amplification).

particle presence was detected through the changes in the sensor resistance (or voltage).

1.2.3 Microfluidic Channel Fabrication

The microchannels (Figure 1.3b) were fabricated in PDMS, with 100×50 μm (length×height), following the method described in the work of Fernandes et al. (2014).

The integration of the MR chip with the PDMS microchannels was achieved through irreversible bonding of the Si_3N_4 and PDMS surfaces. Both surfaces were exposed to ultraviolet/ozone (UVO Cleaner, Jelight, Irvine, CA) for 15 min and then mounted face to face and manually aligned and kept at room temperature (RT) overnight. The ensemble was then mounted in a PCB, where the sensors were wire bonded and the wires protected with silicone (Figure 1.3a).

The raw milk sample's constant flowing through the microchannel's section of 50 μm in height and 100 μm in length was challenging because of its density and colloidal behavior. A surfactant addition to the milk sample, namely Tween 20, was used to achieve a higher dispersion of fat globules (0.1–10 μm), allowing lower interfacial tension and its dimension reduction (Walstra et al., 2006). On the other hand, we adopted the milk preparation method of dairy industries for milk homogenization, using agitation (vortex) and higher temperatures (60°C) to decrease fat globules dimension and allow its uniform distribution in the sample.

1.2.4 Readout Electronics

The multichannel PCB designed to interface 15 SVs was connected to an amplifier with an operating gain of 5000×, a high-pass and low-pass filter of 300 Hz and 10 kHz, respectively. Each channel included a configurable dc source, from 0.25 to 2 mA (Costa et al., 2014).

In this work, only one sensor per channel was monitored. One syringe pump was attached to the system, and was the only device not operating under dc batteries (thus, introducing the 50 Hz noise from the main power grid). The sensor output signals were recorded over time using a connection to an acquisition setup composed of a 16 bit analog-to-digital converter (ADC board DT9836-12-2-BNC), at 50 kHz acquisition frequency. The resulting digital signals were then post-processed in a software developed in MATLAB, to apply a low-pass digital filter with a cut-off frequency of 2 kHz, allowing real-time noise characterization and data storing in the hard drive for further analysis (Figure 1.4). The sensors used for this work showed a noise level of 2.5–4 μV (peak-to-peak). During the experiments, the pump operation increased the noise level to 3–4.5 μV.

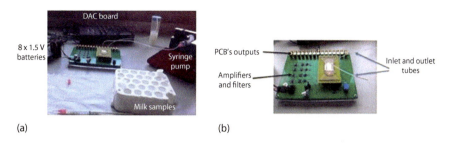

FIGURE 1.4
(a) Acquisition setup assembly and (b) multichannel PCB connected to external electronics.

1.2.5 Biological Functionalization of Nanoparticles

This dynamic detection approach is based on the detection of the fringe field created by magnetic particles attached to the bacterial cells. By selecting suitable antibodies (Ab) (or immunoglobulins [Ig]), it is possible to perform immunological recognition of the targeted mastitis pathogens. The specificity of selected antibodies was tested and proved by immunoblotting (data not shown).

One general differentiation of the streptococci is the Lancefield groups based on serological grouping determined by the antigen C-substance that is a group-specific cell wall polysaccharide. *S. agalactiae* belongs to Lancefield Group B (GB). A polyclonal anti-GB streptococci immunoglobulin G (IgG; 8435-2000, AbDSerotec), one monoclonal anti-*S. agalactiae* immunoglobulin M (IgM; MA1-10871, Thermo Fisher), a rabbit polyclonal IgG to *S. aureus* ScpA protein (ab 92983, Abcam), and a mouse monoclonal IgM anti-*S. aureus* ATCC 29740 (MCA 5793, AbDSerotec—which also recognizes *S. epidermidis*), were used separately. The antibodies were expected to attach to protein A of Nanomag®-D-spio 50 nm particles (79-20-501, MicromodPartikeltechnologie GmbH) by the Fc fraction (fragment crystallizable region) in IgG and by the joining chain (J chain) in IgM (Figure 1.5). Antibodies and bacterial cells dimensions are shown in Figure 1.5.

The bio-functionalization of NP was achieved by the addition of 7.27 μL from the NP original vial (with 5.5×10^{13} NP per mL) to 0.53 μL of polyclonal anti-GB streptococci antibody (1 mg/mL) (or to 5.5 μL of monoclonal anti-*S. agalactiae* antibody [0.5 mg/mL]) in 492.2 μL (or 487.2 μL) of phosphate buffered saline buffer (PBS), or to 1.08 μL of pAb anti-*S. aureus* (0.5 mg/mL) (or to 2.65 μL of mAb anti-*Staphylococcus* spp. [1 mg/mL]) in 492.2 μL of PBS (or in 490.08 μL for mAb), respectively. The incubation step required 1 h RT and continuous agitation. Final functionalized NP were magnetically isolated by magnetic separation (MS) column (130-042-201 Miltenyi) according to MACS MiltenyiBiotec protocol and eluted with PBS + 0.5% bovine serum albumin (BSA) + 2 mM ethylenediaminetetraacetic acid (EDTA) buffer after removal of the MS column from the magnet. A volume of 2 μL of this final suspension

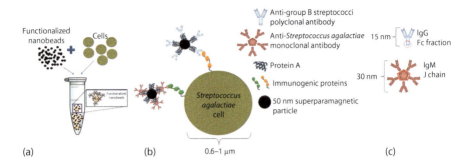

FIGURE 1.5
Schematics of immunomagnetic detection of cells. (a) Incubation of functionalized beads with bacterial cells; (b) biological affinities between different functionalized nanoparticles with bacterial cell wall immunogenic proteins; (c) predictable protein A binding site to each antibody. (Adapted from Duarte, C.M. et al., *Biosensors*, 6(2), 19, 2016.)

with 8×10^6 functionalized NP was diluted in 98 μL of phosphate buffered saline Tween (PBST) and added to each milk sample with bacteria.

1.2.6 Biosensor Calibration

A blank sample (only PBS or sterile raw milk) and a negative control sample (PBS or sterile raw milk), with 2 μL of functionalized NP were always measured prior to the measurements with contaminated milk, giving the background signal of the system.

A first calibration assay was then made with *S. agalactiae*/pAb anti-GB streptococci spiked on a PBS sample.

Finally, *S. agalactiae*/pAb anti-GB streptococci, *S. uberis*/pAb anti-GB streptococci, and *S. agalactiae*/mAb anti-*S. agalactiae* were spiked in sterile milk samples and the correspondent calibration curves made. Each concentration point was the result of three different assays' measurements.

The calibration range between 0.1 and 20 CFU/μL was established taking into account the detection limit for conventional microbiology of 500 CFU/mL (0.5 CFU/μL).

1.2.7 Bacterial Cells

S. agalactiae (strain CECT 183) and *S. uberis* (strain CECT 994) cells were grown separately onto Columbia agar supplemented with 5% sheep blood (bioMérieux, 43021) and incubated at 37°C overnight. A single colony of each isolate was selected and resuspended in 4 mL of Trypticase soy broth over 24 h at 37°C. Subsequently, the bacterial cells were collected through centrifugation (15 min, 17°C, 2700 rpm) and resuspended in PBS 1X (pH 7.2) to allow optical density measurement (at 600 nm) (BECKMAN DU-68 Spectrophotometer) and for colony-forming unit (CFU) estimation. A bacterial suspension with a known concentration of 10^4 CFU/μL was the starting point to get seven different bacterial concentrations for each species, in PBS or in raw milk samples: 0.1, 0.3, 0.5, 1, 2, 10, and 20 CFU/μL.

1.2.8 Sterile Milk Samples

Raw milk used for the definition of calibration curves experiments was collected aseptically from healthy cows. Conventional microbiological tests were performed according to the National Mastitis Council protocols (NMC, 1999), to confirm no bacterial growth. Briefly, a raw milk sample (10 μL) was plated on a Columbia agar plate and on a MacConkey agar plate (CM0007, Oxoid) and both were incubated at 37°C for 48 h. The absence of growth on both plates was considered to be equivalent to the presence of no viable bacteria in the milk.

Each sample for biosensor testing had a 500 μL volume consisting of 2 μL of a suspension of functionalized NP, 98 μL of PBST, and 400 μL of one of seven bacterial suspensions with predefined bacterial concentrations in PBS or sterile raw milk. The incubation of these samples was performed at RT for 3 h under agitation.

All raw milk samples were submitted to a pretreatment of 15 min at 60°C in a dry bath incubator (Grant, model QBD2) and 15 min of continuous centrifugation in a vortex mixer (Labnet) after adding bacteria and PBST. Only then, 2 μL of functionalized NP suspension was added for the final incubation step.

1.2.9 Mastitic Milk Samples

Mastitic milk samples were needed to validate biosensor detection. In these instances, milk samples were collected aseptically from the cows ($n=81$) of 12 Portuguese dairy farms. Animal selection was based on SCC higher than 1,000,000 cells/mL and quarter selection based on CMT results with a score of 3. The quarter was the experimental unit considered ($n=95$). Bacteriological identification was performed as previously mentioned. Milk sample selection was based on bacteriological results that included *S. aureus* ($n=10$), *S. agalactiae* ($n=16$), *S. uberis* ($n=12$), *Streptococcus* spp. ($n=5$), coagulase-negative staphylococci (CNS) ($n=13$), *Enterococcus* spp. ($n=11$), *Escherichia coli* ($n=10$), yeasts ($n=9$), and *Prototheca* ($n=9$). All mastitic milk samples were also submitted for PCR analysis.

1.2.10 PCR Reference Method Analysis

PCR analysis was performed by an external laboratory (VACUNEK, SL) with the PathoProof Mastitis Complete-16® (Thermo Scientific), which allows for the detection of 16 bovine mastitis pathogens. This method is semi-quantitative, classifying the amount of bacterial cells in mastitic milk samples as "high," "average," or "low."

1.2.11 Biosensor Analysis

A volume of 400 µL of mastitic milk was collected and mixed with 98 µL of PBST. Each 498 µL sample was submitted to a pretreatment of heating (15 min at 60°C) and homogenization (15 min in vortex). After these steps, a volume of 2 µL of a functionalized NP suspension was added to reach a final volume of 500 µL to be submitted for incubation (RT, 3 h, under agitation) and further biosensor analysis.

The magnetic detection validation was performed with 95 different mastitic milk samples, consisting on 124 independent trials with four specific antibodies (four groups of 31 milk samples analyzed by specific antibodies as shown in Figure 1.6).

Each assay began with a noise level measurement (Figure 1.7) and each sample was injected at a flow rate of 50 µL/min, through a PDMS microchannel (Figure 1.7). The microchannel was always cleaned between samples with PBST followed by deionized water, both at a 90 µL/min flow rate, until reaching noise level values again, denoting a magnetic-free microchannel filling. A blank sample and a negative control sample were measured whenever a new SV sensor was used and always before mastitic samples analysis.

Samples with functionalized NP on PBS or sterile milk (negative controls) evidenced signals less than 50 µV (Figures 1.8a and 1.9a). Mastitic milk

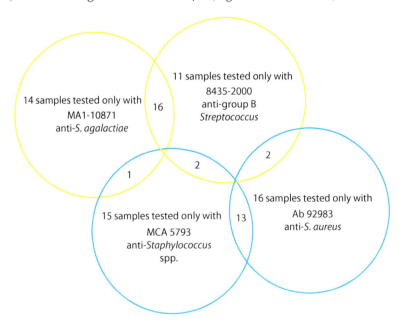

FIGURE 1.6
Number of mastitic samples analyzed per antibody. The interception numbers correspond to the common samples analyzed by respective antibodies. (Adapted from Duarte, C.M., *Portable "Lab-on-Chip" Platform for Bovine Mastitis Diagnosis in Raw Milk*. Lisbon, Faculdade de Medicina Veterinária: Universidade de Lisboa, 2016.)

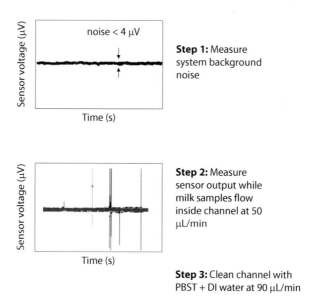

FIGURE 1.7
Biosensor analysis procedure steps.

samples without the targeted bacteria (proved by PCR) and tested with NP functionalized with chosen antibodies, also evidenced signals less than 50 µV. Samples used for calibration assays spiked with bacterial cells on sterile milk showed magnetic signals greater than 50 µV (Figures 1.8b and 1.9b).

The classification of mastitic milk samples by the biosensor was based on bacterial detection (presence or absence). Consequently, the "Positive" samples were those with at least one magnetic peak above 50 µV, therefore higher than the signal found in negative control samples and in mastitic

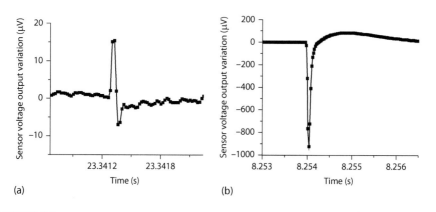

FIGURE 1.8
Sensor output for (a) negative control with the higher amplitude of 23 µV, and the highest amplitude peak of (b) 923.7 µV was obtained in raw milk with anti-*S. agalactiae* and 0.3 CFU/µL of *S. agalactiae*. (Adapted from Duarte, C.M. et al., *Biosensors*, 6(2), 19, 2016.)

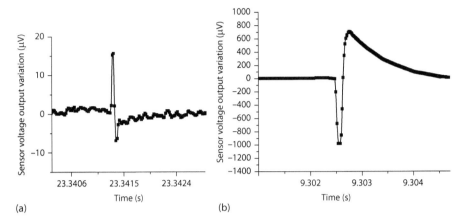

FIGURE 1.9
Sensor output for (a) negative control with the higher amplitude of 23 µV, and the highest amplitude peak of (b) 1703.4 µV was obtained in raw milk with pAb anti-*S. aureus* and 10 CFU/µL of *S. aureus*. (Adapted from Duarte, C.M. et al., *Journal of Dairy Research*, 23, 1–9, 2016.)

milk samples without targeted bacteria. Next, this "Positive" sample magnetic peak evidenced a bipolar or unipolar shape similar to those found in samples used for calibration trials, as shown in Figures 1.8b and 1.9b.

Biosensor analysis was a dynamic detection where a heterogeneous milk sample flowed inside the microchannel. Magnetically labeled bacterial cells were randomly mixed in milk leading to the impossibility of predicting its position above the sensor over time. Consequently, the magnetic peaks shape and time resulting from biosensor analysis were expected to be different between samples (Figures 1.8 and 1.9).

1.2.12 Data Analysis

Isolates were considered to be correctly identified by magnetic detection if the same species was found by the reference method, or if the magnetic detection did not identify the species it was targeting and PCR identified one of the other bacteria. For example, a correct identification referred to *S. agalactiae* being identified magnetically in a sample that PCR had identified as *S. agalactiae*, but also when not identifying as *S. agalactiae* a sample that through PCR was identified as *Staphylococcus* spp. Regarding conventional microbiology, isolates were considered to be correctly identified if the same species was found as with the reference method. Misidentification was considered when the magnetic detection and conventional microbiology identified a different species than the reference method. For example, a sample that was identified as *Staphylococcus* spp. by PCR and that was identified as *S. agalactiae* by magnetic detection or a sample that was identified as *S. agalactiae* by PCR and not identified as such by magnetic detection. For both magnetic detection and conventional microbiology methods, sensitivity, specificity, and positive predictive value were calculated in comparison with PCR species identification. Sensitivity was calculated as the proportion of true positive isolates that were correctly identified with magnetic detection or microbiological tests, for example, the proportion of *S. agalactiae* isolates based on PCR analysis that were identified as such by magnetic detection and microbiology testing. Specificity was calculated as the proportion of true negatives that were correctly identified with magnetic detection and microbiological tests, for example, the proportion of isolates other than *S. agalactiae* based on PCR analysis that were identified as something other than *S. agalactiae* by magnetic detection and by microbiological testing. Finally, PPV was calculated as the proportion of isolates identified as a specific species based on magnetic detection or on microbiological testing that truly represented that particular species, for example, the proportion of isolates that were identified as *S. agalactiae* by magnetic detection or microbiological testing that had been identified as *S. agalactiae* based on PCR analysis.

1.3 Results

1.3.1 Evaluation of Biosensor's Bacterial Quantification

The outcome of the calibration trials are shown in Figures 1.10 and 1.11 for *Streptococcus* and *Staphylococcus* species, respectively. The peak's number per signal amplitude was calculated evidencing no linear correlation with increasing bacterial concentration, in both calibration sets. The milk samples with the anti-GB streptococci antibody revealed the most exuberant signal with *S. uberis* when compared with the other two bacteria–antibody pairs (Figure 1.10). Only the *S. agalactiae*/pAb anti-GB streptococci pair evidenced no peaks higher than 200 µV. Despite that, these MR sensors could detect *S. agalactiae* and *S. uberis* in milk samples from 0.1 CFU/µL (100 CFU/mL).

The milk samples with functionalized NP with the anti-*Staphylococcus* spp. antibody evidenced the highest peak number with *S. epidermidis* when compared with the other two bacteria–antibody pairs (Figure 1.11). Overall, the biosensor could detect *S. aureus* and *S. epidermidis* in milk samples from 100 CFU/mL.

The calibration curve for PBS samples with bacteria was only obtained for the *S. agalactiae*/pAb anti-GB streptococci pair. It was evidenced that different bacterial concentrations, as in sterile milk, also presented similar amplitude peaks (under 200 µV). Performing experimental data fitting to simulations for cell quantity estimation by peak, we obtained different results depending on the considered functionalized NP number per cell and cells cluster positioning above the MR sensor (height z) (Figure 1.12b and c).

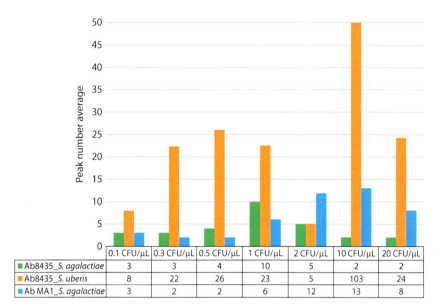

FIGURE 1.10
Calibration trial results for milk samples with seven bacterial concentrations (*S. agalactiae* or *S. uberis*) and functionalized NP with pAb anti-GB streptococci (Ab 8435) and mAb anti-*S. agalactiae* (Ab MA1). The peak number average for each bacteria–antibody pair is counted. (Adapted from Duarte, C.M. et al., *Biosensors*, 6(2), 19, 2016.)

1.3.2 Validation of Magnetic Detection with mAb Anti-*S. agalactiae* and pAb Anti-GB Streptococci

Forty-six mastitic milk samples with known bacteriology results, obtained through conventional microbiology, were analyzed by PCR. A total of 160 identifications were performed by PCR for all 46 milk samples analyzed with mAb anti-*S. agalactiae* and pAb anti-GB streptococci. The most frequently isolated species based on PCR were *Staphylococcus* spp., *E. coli*, and yeasts, followed by *S. uberis* and *S. agalactiae*. As a result of the high sensitivity of the PCR methodology, an average of four different pathogens were detected per mastitic milk sample, not allowing for the true causative agent of mastitis to be determined. Therefore, it was decided to use conventional bacteriology results as the basis for the true identification, confirmed by PCR (Table 1.2).

Magnetic detection with the anti-*S. agalactiae* antibody tested 31 mastitic milk samples from the total of 46 analyzed by conventional microbiology, of which 10 and 13 were identified as *S. agalactiae*, respectively (Table 1.2). However, from these 31 samples tested, 11 were identified as *S. agalactiae* by PCR and only 1 was not identified as such by microbiology, but as *S. uberis*. Magnetic detection with the anti-*S. agalactiae* antibody identified 8 *S. agalactiae* isolates correctly in 11 (72.7%) milk samples with this species when

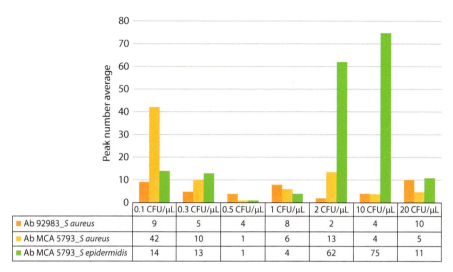

FIGURE 1.11
Calibration trial results for milk samples with seven bacterial concentrations (*S. aureus* or *S. epidermidis*) and functionalized NP with pAb anti-*S. aureus* (Ab 92983) or mAb anti-*Staphylococcus* spp. (Ab MCA 5793). The peak number average for each bacteria–antibody pair is counted. (Adapted from Duarte, C.M. et al., *Journal of Dairy Research*, 23, 1–9, 2016.)

compared to PCR. Five mastitic milk samples did not lead to an identification by this polyclonal antibody because they did not present *S. agalactiae* according to the PCR analysis (true negatives) (Table 1.3). Misidentification was observed for 18 of 31 (58%) isolates in the 31 mastitic milk samples tested with the anti-*S. agalactiae* antibody (Table 1.2). Only three misidentified *S. agalactiae* isolates were found in milk samples with this species analyzed by PCR, evidencing a failure of recognition by this monoclonal antibody (Table 1.2). Adding to that, 5 mastitic milk samples with *S. uberis* and/or *S. dysgalactiae* and 10 mastitic milk samples without any streptococci species according to the PCR analysis were misidentified by biosensor analysis as having *S. agalactiae* and all were classified as false positives (Table 1.3). Overall, 73% sensitivity, 25% specificity, and 35% PPV were found for magnetic detection with the anti-*S. agalactiae* antibody. The highest sensitivity value represents the proportion of true positives (8) that were correctly identified with this monoclonal antibody (Table 1.3).

Using the polyclonal anti-GB streptococci in magnetic detection, the 31 mastitic samples tested included 16 identified equitably as *S. agalactiae* and *S. uberis* by conventional microbiology (Table 1.2). However, PCR analysis identified 2 more samples as *S. uberis* in the 31 analyzed by this antibody, amounting to 18 bacterial target possibilities. Magnetic detection with the anti-GB streptococci antibody correctly identified 7 streptococci isolates present in 18 (38.9%) milk samples with *S. agalactiae* and/or *S. uberis* according to

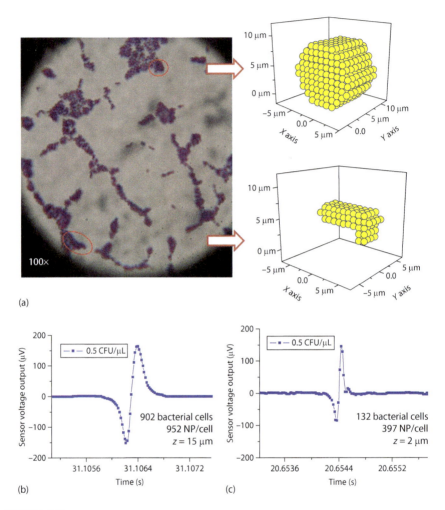

FIGURE 1.12
Microscopic image of *S. agalactiae* cells where a spherical cluster and an elongated cluster are evidenced (a). Experimental data fitting of the highest amplitude peaks obtained in PBS samples with different *S. agalactiae* concentrations (0.5 CFU/μL: 162 μV (b) and 10 CFU/μL: 146 μV (c)) during calibration curve settlement. (Adapted from Duarte, C.M. et al., *Biosensors*, 6(2), 19, 2016.)

PCR analysis. The microorganisms that were not identified as GB streptococci or *S. uberis* (13/31) by the reference method in mastitic milk samples were magnetically detected as GB streptococci and/or *S. uberis* in those samples (5 false positives) or else undetected as true negatives (8) (Table 1.3). Misidentification was observed for 16 isolates in the 31 (51.6%) mastitic samples tested. Eleven misidentified *S. agalactiae* and *S. uberis* isolates were found in milk samples analyzed by PCR with these two streptococci, evidencing a

TABLE 1.2
Identification of Isolates in Mastitic Milk Samples with Both Magnetic Detection (mAb anti-*S. agalactiae*; pAb anti-GB streptococci) and Conventional Microbiology, Compared to PCR Analysis as the Reference Method

Mastitic Milk Isolates	n	Magnetic Detection Anti-*S. agalactiae* Correctly Identified n	%	MI[a]	Magnetic Detection Anti-GB Streptococci Correctly Identified n	%	MI	Microbiological Tests Correctly Identified n	%	MI
S. aureus	1	0	0.0	0	1	100.0	0	0	0.0	1
S. agalactiae	13	7	70.0	3	2	25.0	6	13	100.0	0
S. uberis	11	3	33.3	6	5	62.5	3	8	72.7	3
Streptococcus spp.	4	0	0.0	3	2	50.0	2	1	25.0	3
Staphylococcus spp.	2	0	0.0	0	1	50.0	1	2	100.0	0
Enterococcus spp.	7	3	75.0	1	1	33.3	2	0	0.0	7
Escherichia coli	3	0	0.0	1	1	50.0	1	3	100.0	0
Yeasts	3	0	0.0	2	1	100.0	0	2	66.7	1
Prototheca	2	0	0.0	2	1	50.0	1	2	100.0	0
Total	46	13	41.9	18	15	48.4	16	31	67.4	15

Source: Duarte, C.M. et al., *Biosensors*, 6(2), 19, 2016a.
Note: Correctly identified = true positives + true negatives.
[a] MI (Misidentified) = false negatives + false positives.

TABLE 1.3
Sensitivity, Specificity, and Positive Predictive Value of Magnetic Detection and Conventional Microbiology, Using PCR Analysis as the Reference Method

	Magnetic Detection Anti-*S. agalactiae*	Anti-GB Streptococci	Microbiological Tests
True positives	8	7	31
True negatives	5	8	0
False positives	3	10	0
False negatives	15	6	15
Sensitivity	73%	41%	100%
Specificity	25%	57%	–
PPV[a]	35%	54%	67%

Source: Duarte, C.M. et al., *Biosensors*, 6(2), 19, 2016.
[a] PPV = positive predictive value.

failure of recognition by this polyclonal antibody (Table 1.2). Overall, a sensitivity of 41%, a specificity of 57%, and a PPV of 54% were found for magnetic detection with the anti-GB streptococci antibody. The highest specificity value represents the proportion of true negatives (8) that were correctly identified with this polyclonal antibody (Table 1.3).

With regard to microbiological testing for all 46 samples analyzed with mAb anti-*S. agalactiae* and pAb anti-GB streptococci, the highest microorganism identification, in comparison with PCR analysis, was found to be 100% for *S. agalactiae*, *Staphylococcus* spp., *E. coli*, and *Prototheca*, as shown in the percentage data of correct identification in Table 1.2. However, incomplete microbiological identification of 67.4% (31/46) and a misidentification of 32.6% (15/46) were observed (Table 1.2). Microbiological tests evidenced a PPV value of 67% and a sensitivity of 100% to identify mastitis pathogens in milk samples, showing that conventional microbiology correctly identified true negatives (Table 1.3).

1.3.3 Validation of Magnetic Detection with mAb Anti-*Staphylococcus* spp. and pAb Anti-*Staphylococcus aureus*

Forty-nine mastitic milk samples with known bacteriology results, obtained through conventional microbiology, were analyzed by PCR. A total of 123 agent identifications were attained by PCR for all 49 milk samples analyzed with mAb anti-*Staphylococcus* spp. and pAb anti-*S. aureus*. The species detected by PCR were, in decreasing order of detection, *Staphylococcus* spp., *S. aureus*, yeasts, and *E. coli*, followed by *S. uberis* and *Prototheca*. As a result of the high sensitivity of the PCR methodology, an average of three different pathogens were detected per mastitic milk sample, not allowing for the true causative agent of mastitis to be determined. Therefore, it was decided to use conventional bacteriology results as the basis for true identification, confirmed by PCR (Table 1.4).

Magnetic detection with the anti-*S. aureus* antibody tested 31 mastitic milk samples from the total of 46, of which 6 and 9 were identified as *S. aureus* by conventional microbiology, respectively (Table 1.4). However, of these 31 samples tested, 7 were identified as *S. aureus* by PCR and the one not identified as such by microbiology was *Prototheca*. Magnetic detection with the anti-*S. aureus* antibody identified four *S. aureus* isolates correctly out of seven (57.1%) milk samples with this species according to PCR results. Eighteen mastitic milk samples did not lead to an identification by this polyclonal antibody as they did not present *S. aureus* according to the PCR, thus being true negatives (Table 1.5). Misidentification was observed for 9 of 31 (29%) isolates in all mastitic milk samples tested with the anti-*S. aureus* antibody (Table 1.4). Only three misidentified *S. aureus* isolates were found in milk samples with this species analyzed by PCR, evidencing a failure of recognition by this monoclonal antibody (Table 1.4). Adding to that, the remaining six mastitic milk samples were misidentified as having *S. aureus* (Table 1.4), while really presenting *S. uberis*, yeasts, *S. agalactiae*, and other *Staphylococcal* species according to PCR. Overall, 57.1% sensitivity, 75% specificity, and 40% PPV were found for magnetic detection with the anti-*S. aureus* ScpA antibody. The highest specificity value represents the proportion of true negatives (18) that were correctly identified with this monoclonal antibody (Table 1.5).

TABLE 1.4

Identification of Isolates in Mastitic Milk Samples with Both Magnetic Detection (pAb anti-*S. aureus*; mAb anti-*Staphylococcus* spp.) and Conventional Microbiology, Compared to PCR Analysis as the Reference Method

		Magnetic Detection						Microbiological Tests		
		Anti-*S. aureus*			Anti-*Staphylococcus* spp.					
		Correctly Identified			Correctly Identified			Correctly Identified		
Mastitic Milk Isolates	n	n	%	MI[a]	n	%	MI	n	%	MI
S. aureus	9	3	50.0	3	5	83.3	1	9	100.0	0
Staphylococcus spp.	11	6	75.0	2	5	71.4	2	10	90.9	1
S. agalactiae	3	2	66.7	1	3	100.0	0	3	100.0	0
S. uberis	1	1	100.0	0	1	100.0	0	1	100.0	0
Streptococcus spp.	1	0	0.0	1	0	0.0	0	0	0.0	1
Enterococcus spp.	4	1	33.3	2	1	100.0	0	0	0.0	4
E. coli	7	2	100.0	0	4	80.0	1	7	100.0	0
Yeasts	6	4	100.0	0	3	100.0	0	4	66.7	2
Prototheca	7	3	100.0	0	2	40.0	3	7	100.0	0
Total	49	22	71.0	9	24	77.4	7	41	83.7	8

Source: Duarte, C.M. et al., *Journal of Dairy Research*, 23, 1–9, 2016.
Note: Correctly identified = true positives + true negatives.
[a] MI (misidentified) = false negatives + false positives.

TABLE 1.5

Sensitivity, Specificity, and Positive Predictive Value of Magnetic Detection and Conventional Microbiology, Using PCR Analysis as the Reference Method

	Magnetic Detection		Microbiological Tests
	pAb Anti-*S. aureus*	mAb Anti-*Staphylococcal* spp.	
True positives	4	23	41
True negatives	18	1	0
False positives	3	6	1
False negatives	6	1	7
Sensitivity	57.1%	79.3%	97.6%
Specificity	75.0%	50.0%	—
PPV[a]	40.0%	95.8%	85.4%

Source: Duarte, C.M. et al., *Journal of Dairy Research*, 23, 1–9, 2016b.
[a] PPV = positive predictive value.

Using the monoclonal anti-*Staphylococcus* spp. antibody in magnetic detection, the 31 mastitic samples tested included 6 *S. aureus* and 7 *Staphylococcus* spp. identified by conventional microbiology (Table 1.4). However, PCR analysis identified 29 as *Staphylococcus* spp. of which 8 samples also evidenced *S. aureus*, showing incomplete identification by microbiological analysis. Magnetic detection with the anti-*Staphylococcus* spp. antibody correctly identified 23 staphylococci present in 29 (79.3%) milk samples with mostly staphylococci other than *S. aureus*, or also with *S. aureus* according to PCR analysis. Only one mastitic milk sample was not detected by this monoclonal antibody (true negative) because it did not present any *Staphylococcus* spp. according to PCR (Table 1.5). Misidentification was observed for 7 isolates out of the 31 (22.6%) mastitic samples tested. Six isolates were found in milk samples with *Staphylococcal* species analyzed by PCR, evidencing a failure of recognition by this anti-*Staphylococcus* spp. antibody (Table 1.4). Moreover, only one false positive was found in a mastitic sample without any staphylococci evidenced by PCR. Overall, a sensitivity of 79.3%, a specificity of 50%, and a PPV of 95.8% were found for magnetic detection with the anti-*Staphylococcus* spp. antibody (Table 1.5).

Regarding microbiological testing for all 49 samples considered, 100% of *S. aureus*, *S. agalactiae*, *S. uberis*, *E. coli*, and *Prototheca* were correctly identified when comparing with PCR (Table 1.4). Incomplete microbiological identification of 85.7% (42/49) and a misidentification of 14.3% (7/49) were observed. Microbiological tests evidenced a PPV value of 85.4% and a sensitivity of 97.6% to identify mastitis pathogens in milk samples, showing that conventional microbiology correctly identified true positives (41/49) (Table 1.5).

1.4 Discussion

Sensitivities of 73% and 41% and specificity values of 25% and 57% were obtained for the magnetic identification of streptococci species with an anti-*S. agalactiae* antibody and an anti-GB streptococci antibody, respectively. With regard to the magnetic identification of staphylococci species in mastitic milk samples with an anti-*S. aureus* ScpA antibody and an anti-*Staphylococcus* spp. antibody, sensitivities of 57.1% and 79.3% and specificity values of 75% and 50% were obtained, respectively.

The knowledge of *S. aureus* as an important cause of udder infections in dairy herds sustains the greatest interest in treatment and prevention studies of *S. aureus* mastitis (Fabres-Klein et al., 2014). So, comparing the sensitivity and specificity values of this magnetic detection with another study that used immunological detection of mastitis pathogens through an enzyme-linked immunosorbent assay (ELISA) for detecting *S. aureus* in milk, a sensitivity of between 69% and 90% and specificity values of 61%–97% (Hicks et al., 1994) were observed. That test had a detection limit of 10^4–10^5 CFU/mL, when the minimum bacterial presence detected by the present immunological recognition was 100 CFU/mL, independently of antibody and targeted bacteria. Adding to that, a sandwich ELISA test recently patented to detect *S. aureus* in artificially contaminated milk (Libing et al., 2012), found a similar detection limit of 10^5 CFU/mL. However, a competitive immunoassay performed by an amperometric magnetoimmunosensor (de Ávila et al., 2012) for the specific detection and quantification of *Staphylococcal* protein A and *S. aureus* cells, evidenced a detection limit of 1 CFU/mL, also in artificially contaminated milk samples.

Comparing again the sensitivity (41% and 73%; 57.1% and 79.3%) and specificity (25% and 57%; 50% and 75%) values obtained for the magnetic identification of streptococci and staphylococci species, respectively, with another study based on immuno-agglutination, which compared six commercially available slide agglutination tests for *S. aureus* identification in milk samples (Zschöck et al., 2005), the highest sensitivity (86.7%) and specificity (90.1%) values were obtained for a test consisting of latex particles coated with human fibrinogen and IgG. Still, strain typing methods that are DNA sequence based have also been used to improve *S. aureus* detection. The Bittar et al. (2009) study to differentiate between positive and negative *S. aureus* strains for Panton–Valentine leukocidin, used matrix-assisted laser desorption ionization–time of flight mass spectrometry (MALDI-TOF MS) analysis, which evidenced higher sensitivity (100%) and specificity (90.6%) compared with our magnetic detection method.

Another method for the identification of bovine mastitis pathogens resorted to microarray technology, which was capable of detecting seven common species of mastitis-causing pathogens within 6 h, with an observed sensitivity of 94.1% and a specificity of 100% (Lee et al., 2008). The platform used was based on PCR technology where pathogen-specific targets of DNA were amplified and transferred to react and hybridize with specific probes that were pre-spotted on a biochip. At the end of the process, colorimetric techniques were used to identify pathogen patterns present on the sample. The detection limit of this method was 10^3–10^5 CFU/mL. Despite the advantage of using nucleic acid amplification strategies, which increase the sensitivity, specificity, and efficiency, these methods always required the pre-isolation of bacterial cells from milk, not allowing for the direct analysis of mastitic milk samples and, consequently, their use on-farm, unless the pretreatment step was incorporated inside the analysis system, reducing its time and cost.

According to Lazcka et al. (2007), in order to become attractive, biosensors first need to show that they are capable of reaching at least the same detection levels as traditional techniques (between 10 and 100 CFU/mL). Next, they need to do so in a fraction of time without overlooking cost. Currently, the detection limits of biosensors for on-site use are a hundred to a million cells per milliliter of sample (10^2–10^6 CFU/mL) and are able to achieve extremely high sensitivities (Yoon & Kim, 2012).

Regarding the validation of the magnetic detection method, some false positive results could be explained by NP agglomeration by the mastitic milk matrix heterogeneity, the sporadic low cleaning efficiency of the channel's inlet chamber, and also due to the electrical conductivity of mastitic milk samples. An effect of bovine mastitis is the ion concentration changes in the mammary gland due to increased vascular permeability resulting from an inflammatory response, leading to modifications in the electrical conductivity of milk (Hovinen et al., 2006). The conductance in milk causes a sensor's resistivity variation translated by higher background noise. Despite a detergent (Tween 20) being in milk samples to reduce fat globules dimensions, to distribute the bacterial cells in the milk, to improve NP mobilization, and to allow a more homogeneous milk matrix, the optimization trials (not described in this chapter) showed the need for a compromise between Tween 20 quantity and magnetic peak discrimination. Different volumes of PBST in 500 µL of milk samples were used. The higher volumes (>100 µL) evidenced bubbles inside the microchannel that hampered the milk flow, caused NP agglomeration, and did not help magnetic peaks discrimination between control samples (milk with only NP) and samples with bacteria. On the other hand, false negatives may have occurred because of three circumstances. Firstly, the binding yield variations between antibodies and NP and/or failure in bacterial cells magnetic labeling could narrow bacterial cells identification. This fact should be recognized as possible because IgM and NP dimensions are closer, so it will be more difficult to have the same number of attached IgM when comparing with NP functionalization with smaller IgG. Secondly, according to Henriksen et al.'s (2015) study, it is possible that the rotating nature of the magnetic dipole field of NP magnetized by an external magnetic field can induce signal cancelation. Therefore, the fields from two differently placed NP can partially cancel each other. Finally, microchannel current height (50 µm) could be reduced to improve a sensor's detection by forcing bacterial cells dragging over it, but mastitic milk trials showed that milk clots hamper sample flowing and height decrease leads to microchannel obstruction, pointing to a compromise between the sensor's detection and sample fluidity.

As regards bacterial quantification data, this magnetic detection method showed some microbiological and immunological constraints. Bacterial cells group together randomly depending on growth conditions (Quinn et al., 1994). Each bacteria may express a different number of cell wall proteins, including the immunogenic ones (van der Woude & Baümler, 2004). Together, these facts limit the knowledge of how many immunogenic proteins there are per cell and, consequently, how many proteins will be recognized by each specific antibody used. On the other hand, the chemical and colloidal changes of milk components in a state of inflammatory response (Walstra et al., 2006) as occurs with mastitis, prevent and reduce bacterial magnetic labeling efficacy (Duarte et al., 2016a). Consequently, it was not possible to predict peak profiles (number, shape) for each bacterial concentration.

Although the sensitivity of the magnetic detection method is important, many additional factors must be considered, including rapidity, easy to use, flexibility, portability, and costs (Mortari & Lorenzelli, 2014). This dynamic methodology showed that it was possible for a mastitic milk sample to be processed until a final result was obtained in 5 h, but was not suitable for processing a large number of samples (maximum of 10–12 per day). It also showed technical simplicity when established, and ease of scoring and interpreting the results. Despite the lower sensitivities obtained, both antibodies used were capable of detecting bacterial cells in real milk samples. However, other antibodies could be used for further identification of different bovine mastitis pathogens, reinforcing this method's flexibility.

Therefore, despite the need for improvements in bacteriological infection screening and considering the low values for sensitivity and specificity, the magnetic detection method that the current study describes, may be a tool in the future to complement traditional methods in the identification of some important mastitis pathogens. Data gathered from this work, including the minimum bacterial concentration detected (100 CFU/mL), may provide a useful tool for rapid on-farm diagnosis of mastitis pathogens, contributing to both improving animal health and welfare and rationalizing and reducing the use of antibiotics, with positive effects on the economy of dairy farming and on public health.

1.4.1 Strengths, Weaknesses, Opportunities, and Threats

The two main strengths of this magnetic detection method are the successful evidence of immunological recognition of targeted bacteria by the specific antibodies used, and the NP attachment to selected antibodies on positive milk samples (sterile milk with functionalized NP and spiked with known bacterial concentration) when compared to the control samples (sterile milk with only functionalized NP).

Despite this, bacterial quantification is a limitation as there is a lack of knowledge of how many immunogenic proteins are expressed per cell, and consequently, how many proteins will be recognized by each specific antibody used. Simulations reported in Jitariu et al. (2016) can partially support the experimental results, were helpful for the peak's amplitude and shape interpretation, without the assurance of bacterial cells number.

Loureiro et al.'s (2011) study describes an accurate way to extrapolate cell numbers from a known saturation moment of NP of a sample using a hemocytometer. Unfortunately for us, that model fits well with single human cells from the acute myeloid leukemia cell line, but not with bacterial clusters.

The number of NP attached to bacterial cells due to the interaction between the specific antibody and the immunogenic cell wall protein depends primarily on the number and distribution of the different antigens over the bacterial cell's surface. Even knowing the saturation moment of one NP (2.7×10^{18} Am^2), it was not possible, in our working case, to extrapolate the average number of NP per bacterial cluster because we could not know the total number of cells in it.

Adding to this, neighboring cells to bacterial clusters probably could not be magnetically identified. This second method's weakness is explained by the rotating nature of the magnetic dipole field of NP magnetized by an external magnetic field, which can induce signal cancelation (Henriksen et al., 2015). Therefore, the magnetic fields from two differently placed NP could partially cancel each other.

Nevertheless, we can identify five opportunities to improve the developed magnetic method. The first one is about the biosensor's suitability for routine diagnosis, meaning its applicability to large numbers of milk samples. The current number of analyzed samples is 10–12 per day. The external permanent magnet positioned below the 28 sensors (7 per microchannel) creates a magnetic field that affects its transfer curves and sensitivities, except for one sensor focused by the magnet for correct positioning. Thus, only one microchannel can be used, from four available, limiting to one the sample analysis rate. To correct this issue, present research on this topic toward multiplexing includes varying the magnet's type and strengths and also its relative distances to a sensor's PCB (Soares, 2015). However, for the magnet optimal distance found (2 cm), corresponding to unchanged transfer curves for every 28 SVs, an NP size of 130 nm or more was needed to be magnetically detected, because of the weak magnetic fringe field from smaller particles. Further optimization trials could be done to confirm bacterial identification in mastitic milk samples with these larger NP (higher volume), but inherent binding yield issues should be expected.

The biosensor's flexibility is the second opportunity for improvement, which includes using other specific antibodies for further identification of other important bovine mastitis pathogens.

Thirdly, the binding yield variations of NP functionalization could also be enhanced. The difficulty of having the same number of attached IgM (30 nm) to an NP (50 nm) is predictable when compared to smaller IgG (15 nm), which is translated by uncontrollable binding yield variation. This issue is justified by both antibodies stereochemistry and NP volume, which are not possible to change. Consequently, it affects bacterial cells magnetic labeling efficiency. However, there are some methods that can be used, such as thermogravimetric analysis, which applied to a functionalized NP solution, will be able to extrapolate the weight for antibodies and NP in a sample and, consequently, to quantify the binding yield. The thermogravimetric method consists of a thermal analysis that changes the physical and chemical properties of materials measured, as a function of increasing temperature (with constant heating rate) or as a function of time (with constant temperature and/or constant mass loss). This method is commonly used to determine the selected characteristics of materials that exhibit either mass loss or gain due to decomposition, oxidation, or loss of volatiles (such as moisture). The inconvenience of this analysis in our case, was the minimum quantity of sample required. Forty milligrams was too much when compared to 364.03 µg in solution per trial day when we used IgG antibodies or 366.25 µg in solution per trial day when we used IgM antibodies, and the cost to obtain such higher quantity of functionalized NP, was also expensive.

The fourth magnetic method's improvement opportunity includes the multichannel PCB electronic troubleshooting. The multichannel PCB was the main component of the magnetic detection device. Consequently, the identification of the causative problem was critical for further correction and continuing daily work. The major problem found was translated visually on signal output oscillations, which prevented further measurements.

Two threats were identified: the first one was based on the sterile and mastitic milk matrix heterogeneity. The first main goal of this project was the achievement of a portable lab-on-chip device, able to be used on-site and to analyze raw milk collected directly from a potentially infected cow. However, milk is a colloid of fat globules and water with dissolved carbohydrates and protein complexes (Walstra, 2006), where bacteria, when present, are distributed throughout the emulsion, suspended in solution as well as entrapped and adsorbed on proteins micelles and fat globules. This knowledge led us to several months of trials with different experimental conditions, including thawing or filtering milk samples (pore diameter >2–3 µm) to remove fat globules from raw milk; redesigning the microchannel's layout including pillars for milk sieving, to help fat globules to disperse; detergent/surfactant concentration (PBST) added to raw milk; temperature variation (4°C or RT) of milk samples with bacterial-labeled cells, before biosensor analysis; different bacterial concentrations spiked into sterile milk to achieve detection limit and NP content optimization in control milk samples for no magnetic signal achievement.

The thawing and filtering steps removed fat globules but also bacterial cells from mastitic milk samples. These hypotheses to improve magnetic detection in raw milk samples were thus abandoned.

We evidenced that we could not get away from a milk sample pretreatment step, even a short one, which should include higher temperature, surfactant addition, and stronger agitation, to reduce fat globule dimensions, to distribute the bacterial cells in the milk, to improve NP mobilization, and to allow for a more homogeneous milk matrix. The 60°C temperature value was tested because it is the same used by the dairy industry for the milk homogenization step, to reduce fat globules dimensions and to allow their uniform distribution in raw milk. Those conditions were expected to allow better access to bacterial cells by functionalized NP and were confirmed by further results.

Finally, the last threat to the biosensor's performance was the fact that MR sensors were affected by the electrical conductivity of mastitic milk.

An effect of bovine mastitis is changes in milk's ion concentrations due to increased vascular permeability leading to modifications in its electrical conductivity (Hovinen et al., 2006). The conductance in milk causes a sensor's resistivity variation translated by higher background noise instead of a true magnetic signal.

Studies taking advantage of the electrical conductivity of ions in a sample (Hassan et al., 2014), or an impedance signal of particles and cells using the surrounding media as a reference (Gawad et al., 2001), are not suitable for mastitic milk samples because these present high sensitivity to the sample matrix and cannot distinguish between the conductance of milk components and bacterial presence, which greatly hinders these devices' use outside laboratory facilities. This knowledge reinforces the better suitability of the magnetic detection method studied and described in this work.

1.5 Future Perspectives

The development prospects for new bovine mastitis diagnosis methodologies point to new biomarkers and technological advances for high sensitivity and specificity, fast and efficient devices that can offer a "cow-side" use.

Giouroudi and Keplinger (2013) outlined that several novel manipulation, separation, and detection mechanisms based on magnetic methods are continuously emerging, proving that magnetic biosensing has the potential to become competitive and probably replace in the future the current optical and fluorescence detection technologies, while maintaining the high sensitivity and fast readout time.

Consequently, the magnetic detection device described in this chapter can also be a part of that future. Taking into account the mentioned strengths and opportunities, this biosensor can be submitted for further improvements, which may include a milk pretreatment step incorporated into the microfluidic platform and also further studies on electronics to allow multiplex analysis of several samples at a time. Currently, this biosensor requires an external computer for system operation and displaying test results, so a fully integrated system into a single device could also be made.

The bacterial quantification, however, may be done within lower/upper threshold limits. Simulations of the sensor output as a function of the NP distribution over the cells (using colonies/clusters configurations compatible with that experimentally observed under a microscope) can provide an indication on minimum and maximum numbers. Further work could be done toward a more accurate quantification based on simulations.

References

Åkerstedt, M., Forsbäcka, L., Larsena, T. & Svennersten-Sjaunja, K. (2011). Natural variation in biomarkers indicating mastitis in healthy cows. *Journal of Dairy Research*, 78, 88–96.

Baibich, M.N., Broto, J.M., Fert, A., Nguyen Van Dau, F., Petroff, F., Etienne, P., Creuzet, G., Friederich, A. & Chazelas, J. (1988). Giant magnetoresistance of (001)Fe/(001) Cr magnetic superlattices. *Physical Review Letters*, 61(21), 2472–2475.

Bexiga, R., Koskinen, M.T., Holopainen J., Carneiro, C., Pereira, H., Ellis, K.A. & Vilela, C.L. (2011). Diagnosis of intramammary infection in samples yielding negative results or minor pathogens in conventional bacterial culturing. *Journal of Dairy Research*, 78, 49–55.

Bittar, F., Ouchenane, Z., Smati, F., Raoult, D. & Rolain, J.M. (2009). MALDI-TOF MS for rapid detection of staphylococcal Panton–Valentine leukocidin. *International Journal of Antimicrobial Agents*, 34, 467–470.

Bradley, A. (2002). Bovine mastitis: An evolving disease. *The Veterinary Journal*, 164(2), 116–128.

Britten, A.M. (2012). The role of diagnostic microbiology in mastitis control programs. *Veterinary Clinics: Food Animal Practice*, 28(2), 187–202.

Chung, T.D. & Kim, H.C. (2007). Recent advances in miniaturized microfluidic flow cytometry for clinical use. *Electrophoresis*, 28(24), 4511–4520.

Costa, T., Piedade, M.S., Germano, J., Amaral, J. & Freitas, P.P. (2014). A neuronal signal detector for biologically generated magnetic fields. *IEEE Transactions on Instrumentation and Measurement*, 63(5), 1171–1180.

Cremonesi, P., Pisoni, G., Severgnini, M., Consolandi, C., Moroni, P., Raschetti, M. & Castiglioni, B. (2009). Pathogen detection in milk samples by ligation detection reaction-mediated universal array method. *Journal of Dairy Science*, 92(7), 3027–3039.

de Ávila, B.E.F., Pedrero, M., Campuzano, S., Escamilla-Gómez, V. & Pingarrón, J.M. (2012). Sensitive and rapid amperometric magnetoimmunosensor for the determination of *Staphylococcus aureus*. *Analytical and Bioanalytical Chemistry*, 403(4), 917–925.

Dieny, B. (1994). Giant magnetoresistance in spin-valve multi-layers. *Journal of Magnetism and Magnetic Materials*, 136(3), 335–359.

Duarte, C.M. (2016c). Portable "lab-on-chip" platform for bovine mastitis diagnosis in raw milk. Physical Doctor degree in Veterinary Sciences. Dissertation in Biomedical and Biological Sciences. Lisbon: Faculdade de Medicina Veterinária—Universidade de Lisboa.

Duarte, C.M., Carneiro, C., Cardoso, S., Freitas, P.P. & Bexiga, R. (2016b). Semi-quantitative method for staphylococci magnetic detection in raw milk. *Journal of Dairy Research*, Dec 23, 1–9.

Duarte, C.M., Costa, T., Carneiro, C., Soares, R., Jitariu, A., Cardoso, S., Piedade, M.S., Bexiga, R. & Freitas, P.P. (2016a). Semi-quantitative method for *Streptococci* magnetic detection in raw milk. *Biosensors*, 6(2), 19.

Fabres-Klein, M.H., Aguilar, A.P., Silva, M.P., Silva, D.M. & Ribon, A.O.B. (2014). Moving towards the immunodiagnosis of staphylococcal intramammary infections. *European Journal of Clinical Microbiology and Infectious Diseases*, 33, 2095–2104.

Fernandes, A.C., Duarte, C.M., Cardoso, F.A., Bexiga, R., Cardoso, S. & Freitas, P.P. (2014). Lab-on-chip cytometry based on magnetoresistive sensors for bacteria detection in milk. *Sensors*, 14, 15496–15524.

Freitas, P.P., Cardoso, F.A., Martins, V.C., Martins, S.A.M., Loureiro, J., Amaral, J., Chaves, R.C. et al. (2012). Spintronic platforms for biomedical applications. *Lab on a Chip*, 12(3), 546–557.

Freitas, P.P., Ferreira, R., Cardoso, S. & Cardoso, F. (2007). Magnetoresistive sensors. *Journal of Physics: Condensed Matter*, 19, 165221.

Gawad, S., Schild, L. & Renaud, Ph. (2001). Micromachined impedance spectroscopy flow cytometer for cell analysis and particle sizing. *Lab on a Chip*, 1, 76–82.

Gehanno, V., Freitas, P.P., Veloso, A., Ferreira, J., Almeida, B., Sousa, J.B., Kling, A., Soares, J.C. & Silva, MF. (1999), Ion beam deposition of Mn-Ir spin valves. *IEEE Transactions on Magnetics*, 35 (5), 4361–4367.

Giouroudi, I. & Keplinger, F. (2013). Microfluidic biosensing systems using magnetic nanoparticles. *International Journal of Molecular Sciences*, 14, 18535–18556.

Gröhn, Y.T., Eicker, S.W., Ducrocq, V. & Hertl, J.A. (1998). Effect of diseases on the culling of Holstein dairy cows in New York state. *Journal of Dairy Science*, 81(4), 966–978.

Grönlund, U., Hultén, C., Eckersall, P.D., Hogarth, C. & Waller, K.P. (2003). Haptoglobin and serum amyloid A in milk and serum during acute and chronic experimentally induced *Staphylococcus aureus* mastitis. *Journal of Dairy Research*, 70(4), 379–386.

Hassan, U., Watkins, N.N., Edwards, C. & Bashir, R. (2014). Flow metering characterization within an electrical cell counting microfluidic device. *Lab on a Chip*, 14(8), 1469–1476.

Henriksen, A.D., Wang, S.X. & Hansen, M.F. (2015). On the importance of sensor height variation for detection of magnetic labels by magnetoresistive sensors. *Scientific Reports*, 5, 12282.

Hicks, C.R., Eberhart, R.J. & Sischo, W.M. (1994). Comparison of microbiologic culture, an enzyme-linked immunosorbent assay, and determination of somatic cell count for diagnosing *Staphylococcus aureus* mastitis in dairy cows. *Journal of the American Veterinary Medical Association*, 204(2), 255–260.

Hortet, P. & Seegers, H. (1998). Calculated milk production losses associated with elevated somatic cell counts: Review and critical discussion. *Veterinary Research*, 29(6), 497–510.

Hovi, M. & Roderick, S. (1999). Mastitis in organic dairy herds: Results of a two year survey. *Proceedings of a Soil Association Conference: Mastitis, the Organic Perspective*. Stoneleigh, 3 September. Reading, UK.

Hovinen, M., Aisla, A.M. & Pyörälä S. (2006). Accuracy and reliability of mastitis detection with electrical conductivity and milk colour measurement in automatic milking. *Acta Agriculturae Scandinavica, Section A: Animal Science*, 56(3–4), 121–127.

Huh, D., Gu, W., Kamotani, Y., Grotberg, J.B. & Takayama, S. (2005). Microfluidics for flow cytometric analysis of cells and particles. *Physiological Measurement*, 26(3), R73–R98.

Jitariu, A., Duarte, C., Cardoso, S., Freitas, P.P. & Chiriac, H. (2016). Theoretical method for the evaluation of bacterial concentration by magnetoresistive cytometry. *IEEE Transactions on Magnetics* (in press, 2016), 10.1109/TMAG.2016.2623675.

Kokkinis, G., Cardoso, S., Keplinger, F. & Giouroudi, I. (2017). Microfluidic platform with integrated GMR sensors for quantification of cancer cells. *Sensors & Actuators: B. Chemical*, 241, 438–445.

Korhonen, H. & Kaartinen, L. (1995). Changes in the composition of milk induced by mastitis. In: Sandholm M., Honkanen-Buzalski T., Kaartinen L., Pyörälä S. (Eds.). *The Bovine Udder and Mastitis* (pp. 76–82). Gummerus, Jyväskylä, Finland.

Lago, A., Godden, S.M., Bey, R., Ruegg, P.L. & Leslie, K. (2011). The selective treatment of clinical mastitis based on on-farm culture results: I. Effects on antibiotic use, milk withholding time, and short-term clinical and bacteriological outcomes. *Journal of Dairy Science*, 94(9), 4441–4456.

Lazcka, O., Del Campo, F.J. & Muñoz, F.X. (2007). Pathogen detection: A perspective of traditional methods and biosensors. *Biosensors and Bioelectronics*, 22(7), 1205–1217.

Lee, K.H., Lee, J.W., Wang, S.W., Liu, L.Y., Lee, M.F., Chuang, S.T., Shy, Y.M., Chang, C.L., Wu, M.C. & Chi, C.H. (2008). Development of a novel biochip for rapid multiplex detection of seven mastitis-causing pathogens in bovine milk samples. *Journal of Veterinary Diagnostic Investigation*, 20, 463–471.

Libing, W., Chuanlai, X., Qianqian, Y., Xiaofang, D., Shanshan, S. & Xun, Z. (2012). Kit for rapid detection of *Staphylococcus aureus* in sample and detection method thereof. Pat. No. CN102323416 (A).

Lippolis, J.D. & Reinhardt, T.A. (2010). Utility, limitations, and promise of proteomics in animal science. *Veterinary Immunology and Immunopathology*, 138(4), 241–251.

Loureiro, J., Andrade, P.Z., Cardoso, S., da Silva, C.L., Cabral, J.M. & Freitas, P.P. (2011). Magnetoresistive chip cytometer. *Lab on a Chip*, 11(13), 2255–2261.

Loureiro, J., Cardoso, S., Cabral, J.M. and Freitas, P.P. (2007). The 52nd Magnetism and Magnetic Materials Conference, Tampa, FL, EUA, Micro Total Analysis System for hematopoietic stem/progenitor cell separation and counting, Proceedings.

Mansor, R., Mullen, W., Albalat, A., Zerefos, P., Mischak, H., Barrett, D.C., Biggs, A. & Eckersall, P.D. (2013). A peptidomic approach to biomarker discovery for bovine mastitis. *Journal of Proteomics*, 85, 89–98.

Matsushita, T., Dinsmore, R.P., Eberhart, R.J., Jones, G.M., McDonald, J.S., Sears, P.M. & Adams, D.S. (1990). Performance studies of an enzyme-linked immunosorbent assay for detecting *Staphylococcus aureus* antibody in bovine milk. *Journal of Veterinary Diagnostic Investigation*, 2(3), 163–166.

McCarron, J.L., Keefe, G.P., McKenna, S.L., Dohoo, I.R. & Poole, D.E. (2009). Laboratory evaluation of 3M Petrifilms and University of Minnesota Bi-plates as potential on-farm tests for clinical mastitis. *Journal of Dairy Science*, 92(10), 2297–2305.

Milner, P., Page, K.L. & Hillerton, J.E. (1997). The effects of early antibiotic treatment following diagnosis of mastitis detected by a change in the electrical conductivity of milk. *Journal of Dairy Science*, 80, 859–863.

Mortari, A. & Lorenzelli, L. (2014). Recent sensing technologies for pathogen detection in milk: A review. *Biosensors and Bioelectronics*, 60(15), 8–21.

Mujika, M., Arana, S., Castaño, E., Tijero, M., Vilares, R., Ruano-López, J.M., Cruz, A., Sainz, L. & Berganza, J. (2009). Magnetoresistive immunosensor for the detection of *Escherichia coli* O157:H7 including a microfluidic network. *Biosensors and Bioelectronics*, 24(5), 1253–1258.

National Mastitis Council (NMC). (1999). Sampling collection and handling. *Laboratory Handbook on Bovine Mastitis*. National Mastitis Council Inc., Verona, WI.

Pyörälä, S. (2003). Indicators of inflammation in the diagnosis of mastitis. *Veterinary Research*, 34, 565–578.

Quinn, P.J., Carter, M.E., Markey, B. & Carter, G.R. (1994). *Veterinary Clinical Microbiology* (pp. 127–136). Mosby Editors/Elsevier, London.

Soares, A.R.S. (2015). Portable lab-on-chip platform for bovine mastitis diagnosis in raw milk. Master degree Dissertation in Biomedical and Biophysics Engineering. Lisbon: Faculdade de Ciências—Universidade de Lisboa.

Taponen, S., Koort, J., Björkroth, J., Saloniemi, H. & Pyörälä, S. (2007). Bovine intramammary infections caused by coagulase-negative staphylococci may persist throughout lactation according to amplified fragment length polymorphism based analysis. *Journal of Dairy Science*, 90(7), 3301–3307.

Taponen, S., Salmikivi, L., Simojoki, H., Koskinen, M.T. & Pyörälä, S. (2009). Real-time polymerase chain reaction-based identification of bacteria in milk samples from bovine clinical mastitis with no growth in conventional culturing. *Journal of Dairy Science*, 92(6), 2610–2617.

Tenhagen, B.A., Koster, G., Wallmann, J. & Heuwieser, W. (2006). Prevalence of mastitis pathogens and their resistance against antimicrobial agents in dairy cows in Brandenburg, Germany. *Journal of Dairy Science*, 89(7), 2542–2551.

van der Woude, M.W. & Bäumler, A.J. (2004). Phase and antigenic variation in bacteria. *Clinical Microbiology Reviews*, 17(3), 581–611.

Walstra, P., Wouters, J. & Geurts, T. (2006). *Dairy Science and Technology.* (2nd Edn.). Taylor & Francis Group, New York.

Wang, S.X. & Li, G. (2008). Advances in giant magnetoresistance biosensors with magnetic nanoparticle tags: Review and outlook. *IEEE Transactions on Magnetics*, 44(7), 1687–1702.

Wolff, A., Perch-Nielsen, I.R., Larsen, U.D., Friis, P., Goranovic, G., Poulsen, C.R., Kutter, J.P. & Telleman, P. (2003). Integrating advanced functionality in a microfabricated high-throughput fluorescent-activated cell sorter. *Lab on a Chip*, 3(1), 22–27.

Yazdankhah, S.P., Hellemann, A.L., Rønningen, K. & Olsen, E. (1998). Rapid and sensitive detection of *Staphylococcus* species in milk by ELISA based on monodisperse magnetic particles. *Veterinary Microbiology*, 62(1), 17–26.

Yoon, J.-Y. & Kim, B. (2012). Lab-on-a-chip pathogen sensors for food safety. *Sensors*, 12, 10713–10741.

Zadoks, R.N., Middleton, J.R., McDougall, S., Katholm, J. & Schukken, Y.H. (2011). Molecular epidemiology of mastitis pathogens of dairy cattle and comparative relevance to humans. *Journal of Mammary Gland Biology and Neoplasia*, 16(4), 357–372.

Zadoks, R.N. & Watts, J.L. (2009). Species identification of coagulase-negative staphylococci: Genotyping is superior to phenotyping. *Veterinary Microbiology*, 134(1–2), 20–28.

Zschöck, M., Nesseler, A. & Sudarwanto, I. (2005). Evaluation of six commercial identification kits for the identification of *Staphylococcus aureus* isolated from bovine mastitis. *Journal of Applied Microbiology*, 98(2), 450–455.

2

Giant (GMR) and Tunnel (TMR) Magnetoresistance Sensors: From Phenomena to Applications

Càndid Reig and María-Dolores Cubells-Beltrán

CONTENTS

2.1 Introduction 36
2.2 Structures and Phenomena 37
    2.2.1 Sandwich 37
    2.2.2 Spin Valves 37
    2.2.3 Magnetic Tunnel Junctions 39
    2.2.4 Other GMR Structures 40
2.3 Devices 40
    2.3.1 Technological Issues 40
        2.3.1.1 Spin Valves 41
        2.3.1.2 Magnetic Tunnel Junctions 41
    2.3.2 Devices Design 41
2.4 Limitations 43
    2.4.1 Range of Application 43
        2.4.1.1 Noise Mechanisms in GMR/TMR Sensors 43
    2.4.2 Hysteresis 46
    2.4.3 Voltage Offset 46
    2.4.4 Temperature Drifts 46
    2.4.5 Bandwidth 48
2.5 Applications 49
    2.5.1 General Purpose Magnetometers (Compass) 49
    2.5.2 Industrial Applications 49
        2.5.2.1 Automotive 49
        2.5.2.2 Space 49
        2.5.2.3 Electric Current Measurement 51
    2.5.3 Non-Destructive Evaluation 54
        2.5.3.1 Magnetic Detection 54
        2.5.3.2 Eddy Current Testing 54

    2.5.4   Bio-Applications 56
        2.5.4.1   Detection of Bioanalytes 57
        2.5.4.2   Monitoring of Magnetic Fluids 58
        2.5.4.3   Biomedical Signal Detection 58
2.6   Conclusions 58
Acknowledgments 58
References 59

ABSTRACT   Solid state magnetic sensors have inherent characteristics that make them potential candidates in a huge range of applications regarding magnetic field sensing. We mention their high level of integration with electronics, their low weight, low cost, and wide bandwidth, among others. Hall effect devices and magnetoresistance sensors are excellent examples. Giant magnetoresistance (GMR) and tunneling magnetoresistance (TMR) sensors, while maintaining these advantages, have demonstrated better performance figures regarding sensitivity and signal-to-noise ratio (SNR). In this way, GMR/TMR sensors have been considered in scenarios requiring sub-nT measurements, with demonstrated success. In this chapter, we will present the fundamental basis of GMR/TMR and we will describe the state-of-the-art use of GMR/TMR sensors.

2.1 Introduction

The giant magnetoresistance (GMR) effect was first described in 1988 by A. Fert (Baibich et al. 1988) and then in 1989 by P. Grunberg (Binasch et al. 1989). In 2007, both were awarded the Nobel Prize in Physics for their contributions (Thompson 2008). Basically, the GMR effect is a significant change in its resistance with an external field at room temperature. It is observed in multilayered structures with ferromagnetic layers separated by a non-magnetic spacer due to the relative orientation of the magnetization vectors.

Initially, GMR structures were used as sensing elements in the read heads of hard drives. In these applications, the magnetoresistance (MR) level shifted, with the influence of the magnetic field generated by the magnetically stored bits, between two limit states: maximum and minimum resistance, as described by

$$MR = \frac{R^{\uparrow\downarrow} - R^{\uparrow\uparrow}}{R^{\uparrow\uparrow}}$$

where:
  MR is the so-called magnetoresistance level
  $R^{\uparrow\downarrow}$ is the (maximum) resistance in the anti-parallel state
  $R^{\uparrow\uparrow}$ is the (minimum) resistance in the parallel state

Nowadays, these multilayered structures can be engineered in such a way that the quiescent state of the structure is obtained with layers having their magnetic moments in a crossed-axis (orthogonal) configuration by means of a particular deposition process or by the application of an external magnetic biasing. In this way, the transfer curve of the device is notably smoothed, and the quiescent working point is placed in a medium state, thereby providing a useful quasi-linear region that can be utilized for analog magnetic sensing applications.

2.2 Structures and Phenomena

GMR phenomena were initially reported on Fe/Cr thin multilayers (Baibich et al. 1988; Binasch et al. 1989). It was demonstrated that the electric current in a magnetic multilayer consisting of a sequence of thin magnetic layers separated by equally thin non-magnetic metallic layers is strongly influenced by the relative orientation of the magnetizations of the magnetic layers (about 50% at 4.2 K). The cause of this giant variation in the resistance is attributed to the scattering of the electrons at the layers' interfaces. In this way, any structure with metal–magnetic interfaces is a candidate to display GMR. Since then, a huge effort has been made to find structures that enhance this effect (MR levels at room temperature above 200% are achieved in modern GMR structures). We will next describe some of these structures.

2.2.1 Sandwich

A sandwich structure is the general name for multilayered structures. They usually consist of two magnetic layers of an Fe–Co–Ni alloy, such as permalloy, separated by a non-magnetic conductive layer, such as Cu (Ranchal et al. 2002). A general scheme is shown in Figure 2.1a. With magnetic films of about 4–6 nm wide and a conductor layer of about 35 nm, magnetic coupling between layers is slightly small. With these configurations, MR levels of about 4%–9% are achieved, with linear ranges of about 50 Oe. The figures of merit of sandwich devices can be improved by continuously repeating the basic structure, thereby creating a multilayered system. Successful applications of sandwich structures in magnetic field sensing include bio-electronics (Mujika et al. 2009) and angle sensors (Lopez-Martin and Carlosena 2009).

2.2.2 Spin Valves

The origin of spin valves (SVs) comes from the sandwich structure. In SVs, an additional antiferromagnetic (pinning) layer is added to the top or bottom part of the structure, as shown in Figure 2.1b. In this sort of structure, there is

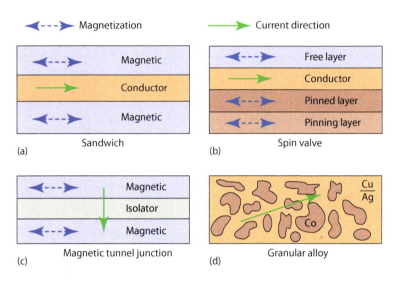

FIGURE 2.1
Basic structures displaying GMR phenomena.

no need for an external excitation to get the anti-parallel alignment. Despite this, the pinned direction (easy axis) is usually fixed by raising the temperature above the knee temperature (at which the antiferromagnetic coupling disappears) and then cooling it within a fixing magnetic field. Obviously, devices so obtained have a temperature limitation below the knee temperature. Typical values displayed by SVs are an MR of 4%–20% with saturation fields of 0.8–6 kA/m (Freitas et al. 2007).

For linear applications, and without excitation, pinned (easy axis) and free layers are arranged in a crossed-axis configuration (at 90°), as detailed in Figure 2.2a. The response of this structure is given by (Freitas et al. 2007)

$$\Delta R = \frac{1}{2}\left(\frac{\Delta R}{R}\right) R_{sq} i \frac{W}{h} \cos(\theta_p - \theta_f)$$

where:
- $(\Delta R/R)$ is the maximum MR level (5%–20%)
- R_{sq} is the sensor sheet resistance (15–20 Ω/sq)
- L is the length of the element
- W is its width
- h is the thickness
- i is the sensor current
- θ_p and θ_f are the angle of the magnetization angle of the pinned and free layers, respectively

Assuming uniform magnetization for the free and pinned layers, for a linearized output, $\theta_p = 90°$ and $\theta_f = 0$.

Giant and Tunnel MR Sensors

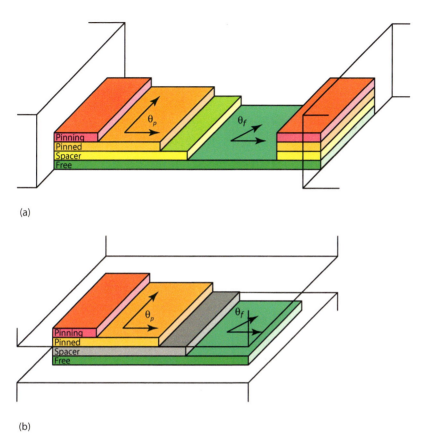

FIGURE 2.2
Multilayer structures corresponding to typical spin valves (a) and magnetic tunnel junctions (b).

2.2.3 Magnetic Tunnel Junctions

Magnetic tunnel junctions (MTJ), also called tunnel magnetoresistance (TMR) structures, were initially described as GMR structures (Hirota et al. 2002). Nowadays, they are considered a specific MR effect (Reig et al. 2013). Nevertheless, due to the similarity of both families of structures and their shared applications, we will also consider them in this chapter.

In this case, the magnetic layers are separated not by a conductive layer but by a very thin isolating layer, following a current perpendicular to planes (CPP) configuration (see Figures 2.1c and 2.2b). Electrons can cross this thin film by means of the quantum tunnel effect. As deduced from quantum mechanics arguments, the crossing probability is higher when both magnetic moments are aligned in parallel and lower when both magnetic moments are not aligned in parallel. The equation describing the output of these structures is

$$\Delta V = \frac{1}{2} \text{TMR}\, i \frac{R \times A}{Wh} \cos(\theta_p - \theta_f)$$

where:
- TMR is the maximum MR level
- i is the biasing current
- $R \times A$ is the resistance per area parameter
- Wh are the dimensions

These devices usually make use of the SV principle in order to fix the easy axis by means of a pinning antiferromagnetic layer. Typical MR levels of MTJ are above 40%, with Al_2O_3 as the isolating layer (Ziese and Thornton 2001). More recently, MR levels of about 200% have been reported for MgO-based structures (Ferreira et al. 2006). Saturation fields are of the order of 1–100 Oe. The basis of linear MTJs is analogous to that of a linear SV. When configured in a crossed-axis configuration, linear ranges suitable for sensor applications can be achieved (Freitas et al. 2007).

2.2.4 Other GMR Structures

Granular films of Co–Cu and Co–Ag also exhibit a GMR effect. In this case, the GMR effect is due to the spin-dependent scattering taking place at the boundaries of Co clusters embedded in the host lattice, as depicted in Figure 2.1d. Because these binary systems are not miscible, the characteristics of the devices are highly conditioned by the growth conditions and the post-deposition treatments. In fact, the amount of MR is accepted to be associated with the size of the Co clusters (Andrés et al. 1999).

GMR can also be found in other structures. We collected two illustrative examples. Pena et al. (2005) report on GMR in ferromagnet/superconductor superlattices and Pullini et al. (2007) describe GMR in multilayered nanowires. In any case, a magnetic–non-magnetic interface is required in order to allow the spin-electron scattering to produce the effect.

2.3 Devices

In order to have functional devices, GMR/TMR multilayered structures have to be patterned into elements with proper resistance values for use as sensors. Then, these elements need to be contacted.

2.3.1 Technological Issues

The deposition of these structures can be done with low temperature processes and then patterned by selective physical etching, so avoiding damage in the substrate. In this sense, the deposition of these structures can be accomplished by ion beam deposition (IBD) or by sputtering. In any case, the substrate temperature does not exceed 120°C. Thus, both processes can be directly masked with photoresist without damaging the substrate. In some cases, a final heat treatment of between 200°C and 300°C is required to increase the MR ratio (He et al. 2010).

2.3.1.1 Spin Valves

A typical structure can be found in Reig et al. (2004). It was deposited by ion beam sputtering (IBD) onto a 3" Si/SiO_2 1500 Å substrate. The SV structure was Ta (20 Å)/NiFe (30 Å)/CoFe (20 Å)/Cu (22 Å)/CoFe (25 Å)/MnIr (60 Å)/Ta (40 Å). This structure was demonstrated to give MR responses of about 6%–7%, linear ranges of about 20 Oe, and sheet resistivities of about 10–15 Ω/sq. Deposition rates ranged from 0.3 to 0.6 Å/s. A 40 Oe field was applied to the substrates during the deposition step in order to state the easy axis in the pinned and free layers. The wafer was rotated 90° between both depositions to ensure a crossed-axis SV configuration.

Nano-oxide layers (NOL) inserted in the pinned layer and above the free layer have been found to increase the MR ratio (Reig et al. 2005). The enhancement of GMR is attributed to the specular scattering effect of the conduction electrons at the metal–insulator interfaces.

2.3.1.2 Magnetic Tunnel Junctions

A typical MTJ structure was deposited by ion beam sputtering (IBD) onto 3" Si/SiO_2 1000 Å substrates. The final structure of the MTJ was Al (600 Å)/Ta (90 Å)/NiFe (70 Å)/MnIr (250 Å)/CoFe (50 Å)/Al_2O_3 (12 Å)/CoFe (50 Å)/NiFe (25 Å)/Ta (60 Å)/TiW (300 Å), as described in Reig et al. (2008). This particular structure was demonstrated to give MR responses close to 40% while keeping linear ranges above 20 Oe.

2.3.2 Devices Design

To implement an SV-based device, only one lithographic step is required for patterning the structures (L1), and then another to design the contacts (L2), at the ends of the SV strip, as shown in Figure 2.3 (left). The sheet resistance is inherent to the specific SV structure, but the final resistance value can be tuned by properly setting L and W of the strip. Usually, the minimum W value is constrained by the lithography resolution, and then the L value is obtained. For SV structures such as those described in the Section 2.3.1.1, devices of 200 × 3 μm give nominal resistances of the order of 1 kΩ.

For MTJs, due to their CPP nature, two masks are required for defining an elemental device. In the first step, a mesa structure is defined (L1) (Figure 2.3, right). Then, a second mask (L2) is applied to define the pillars comprising

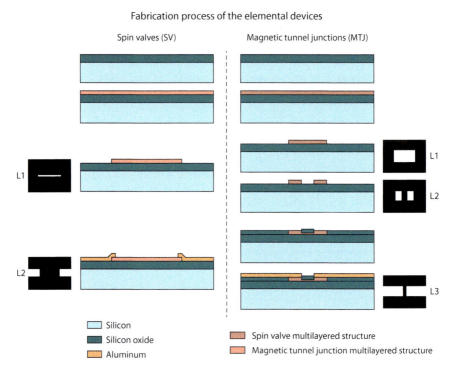

FIGURE 2.3
GMR/TMR basic fabrication steps.

the active region. Usually, a conductive bottom layer is included in the multilayered structure to connect devices in pairs, and facilitating the electrical contacts on the top of the structures (L3).

Because these elements are resistance, once they have been defined, they can be arranged in different configurations, depending on the specific requirement. MTJs are commonly arranged in arrays of elements in series and parallel, due to their better SNR performance and the intrinsic dependence of the MR level with the voltage bias (Chen and Freitas 2012).

In linear applications involving resistive sensors, bridge setups are often considered. They display a highly linear response, a better signal level, zeroed output, and high immunity to undesired external effects. Successful examples of the use of GMR/TMR sensors configured as bridges can be found in Cubells-Beltrán et al. (2009), Guerrero et al. (2009), and Le Phan et al. (2005).

When magnetic imaging is the objective, GMR sensors can also be arranged in arrays, as described, for example, by Cardoso et al. (2006), where MTJ elements have been integrated with a thin-film transistor (TFT) diode for improving the addressing process, or in Hall et al. (2013), where 256 pixels arrays with integrated complementary metal-oxide semiconductor (CMOS) circuitry have been demonstrated.

2.4 Limitations

The limitations of use of these sensors arise from very different reasons including the range of application, reproducibility, voltage offset, temperature drifts, or bandwidth constrictions. In the following sections, we describe some of these limitations and we give the direction of current investigations for overcoming them.

2.4.1 Range of Application

The intrinsically useful range of MR-based sensors is mainly limited by two mechanisms. At the low-level region, the SNR ratio sets the detection limit. Then, a study of the related noise sources is mandatory. The noise power spectrum density (PSD) is commonly given in square volts per hertz (V^2/Hz). Often, it is much more convenient to use the amplitude spectrum density (ASD), expressed in volts per square root hertz (V/$\sqrt{Hz}$), for a comparison with voltage signals. The sensitivity for an MR signal, S_V, is usually given in V/V/T. Typical values for GMR sensors are 20–40 V/V/T, for example, 20–40 nV/nT when they are biased with 1 V. For comparing different sensors, it is recommended to use the field equivalent noise power spectra density, sometimes called *detectivity*. It corresponds to the PSD divided by the sensitivity. For example, if a sensor displays a noise of 10 nV/$\sqrt{Hz}$ at a given frequency and a sensitivity of 25 V/V/T, its detectivity will be 400 pT for 1 V bias (Reig et al. 2013). At the high signal region, the saturation field is the limitation mechanism. For sensing applications, a good approximation is to consider the linear range to be one-half of the saturation field.

GMR/TMR can be applied in the range from some petatesles (pT) to almost kilotesles (kT), which is more than 14 orders of magnitude, as observed in Figure 2.4, compared with other magnetic sensors.

2.4.1.1 Noise Mechanisms in GMR/TMR Sensors

2.4.1.1.1 Thermal Noise

The most relevant noise is the thermal noise (also called the Johnson–Nyquist noise or white noise), which is directly related to the resistance of the sensor. It is a white noise, so it is independent of the frequency. It was first observed by Johnson (1928) and interpreted by Nyquist (1928). It is expressed as

$$S_V(\omega) = \sqrt{4Rk_BT}$$

where:
    R is the sensor resistance
    k_B is the Boltzmann constant
    T is the temperature

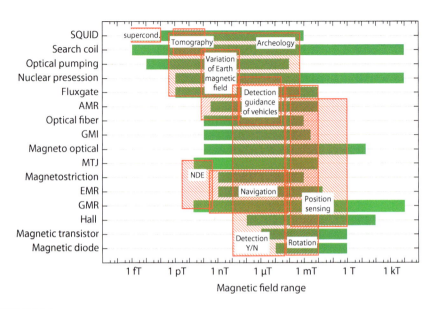

FIGURE 2.4
Range of applications of different magnetic field sensors. (From Díaz-Michelena, M., *Sensors [Basel, Switzerland]*, 9(4), 2271–88, 2009.)

For example, a 1 kΩ resistor at room temperature has 4 nV/√Hz.

2.4.1.1.2 1/f Noise

The origin of the 1/f noise or "pink" noise or Flicker noise is from resistance fluctuations, so it can only be revealed by applying a current to the sensor. Its dependence with the frequency is described by the following phenomenological formula:

$$S_V(\omega) = \frac{\gamma_H R^2 I^2}{N_c f^\beta}$$

where:
- γ_H is a dimensionless constant proposed by Hooge (1976)
- R is the sensor resistance
- I is the bias current
- N_c is the number of current carriers
- f is the frequency
- β is an exponent typically of the order of 1

1/f noise can exhibit a non-magnetic and a magnetic component with possible different slopes. For TMRs, the formula becomes

$$S_V(\omega) = \frac{\alpha R^2 I^2}{A f}$$

Giant and Tunnel MR Sensors

where:

- *A* is the active surface of the device
- α is a parameter with the dimension of a surface

The size and shape of the sensors have a strong effect on the 1/*f* noise.

In GMRs and TMRs, this low-frequency noise is dominant and is often a drawback in the performance of the sensors. Although the TMR in MTJs is significantly higher than the MR in SVs, the intrinsic noise of an MTJ sensor is also higher than that of an SV (by a factor of about 3 [Freitas et al. 2007]). Due to its average nature, small GMR sensors display more 1/*f* noise than bigger sensors. By considering equally thin sensors, the 1/*f* noise is roughly inversely proportional to their area (Reig et al. 2013). As a graphical summary, Figure 2.5 shows the measured noise for several GMR commercial sensors, expressed as detectivity (Stutzke et al. 2005).

2.4.1.1.3 Random Telegraph Noise

The random telegraph noise (RTN; or "popcorn" noise) is due to the fluctuations of a specific source between two different levels with comparable energies and a barrier height able to give a typical characteristic time in the measurement range. RTN is difficult to handle and a sensor with RTN noise is in general very difficult to use even if it is theoretically possible to partially suppress this noise by data treatment.

2.4.1.1.4 Shot Noise

Shot noise (Freitas et al. 2007) arises in discontinuities in the conduction medium as a consequence of the discrete nature of the electrical charge and

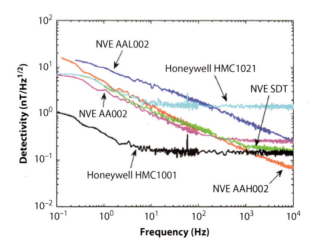

FIGURE 2.5
Low-frequency noise of commercial GMR sensors. (Reprinted with permission from Stutzke, N. A. et al., *Journal of Applied Physics* 97(10), 10Q107, 2005.)

is described by the following equation, where e stands for the electron electrical charge and I is the current flowing through the device (for practical device operating temperatures).

$$S_V(\omega) = 2eIR^2$$

The tunnel barrier of an MTJ is an example of a discontinuity in the conduction medium, which is why shot noise is found in MTJs but not in SVs or anisotropic magnetoresistance (AMR) sensors, which are made of continuous metallic layers.

2.4.2 Hysteresis

The magnetic nature of GMR-based sensors implies an associated hysteresis. Such effect has been analyzed for GMR sensors both numerically (Áč 2008) and experimentally (Liu et al. 2012). Intrinsically, this inherent hysteresis can be internally reduced by considering a so-called crossed-axis configuration of the magnetization moments of the constituent layers of the MR structure. In this case, since the free layer would never reverse, but merely displace by 90°, hysteretic errors below 1% of the full scale can be achieved (Reig et al. 2004). Alternating current (ac) biasing schemes have been proposed for reducing the hysteresis. In Ripka et al. (1999), by using an ac biasing of 5 mA at 10 kHz, the hysteresis of a commercial NVE sensor was reduced from 5% to 1% in the 0.3 mT range. In addition, both SNR and offset were also partially reduced. Hysteresis has also been reduced by biasing the sensors with an external magnetic field (Vopálenský et al. 2004). More recently, electrical models have been developed to reduce the hysteresis in specific GMR-based current sensors (Jedlicska et al. 2010; Han et al. 2015).

2.4.3 Voltage Offset

MR sensors are commonly used in a bridge configuration, so voltage offsets are introduced during the fabrication process. Common sources of these deviations are deposition inhomogeneities and lithography tolerances. As the dimensions of MR structures are close to the lithography resolution limit, these tolerances need to be externally corrected using external circuits.

2.4.4 Temperature Drifts

Temperature is always a limiting parameter in electronics. Every electronic device has a temperature-dependent response arising from its physical nature. Regarding specific GMR sensors, not only does the resistance (and the sensor impedance) vary with the temperature, but so does the MR level (and then the sensitivity).

The resistance of GMR sensors is a function of the temperature. For GMR-based devices, and in the usual range of utilization, this dependence can be considered linear, and can be defined by a temperature coefficient (TEMPCO) as follows:

$$\mathrm{TCR}(\%) = 100 \times \frac{1}{R_{T_0}} \frac{\Delta R}{\Delta T}$$

An analog relationship can be defined for the thermal dependence of sensitivity as

$$\mathrm{TCS}(\%) = 100 \times \frac{1}{S_{T_0}} \frac{\Delta S}{\Delta T}$$

When a full bridge configuration is considered, this thermal dependence is partially compensated and is expected to be low. Due to the inherent voltage offset of sensors configured as bridges, the temperature drift of the offset voltage must be specified:

$$\mathrm{TCV_{off}}(\%) = 100 \times \frac{\frac{\Delta V_{off}}{\Delta T}}{V_{off,T_0}}$$

MR structures are temperature dependent. For real applications, the temperature coefficient of the output voltage of a given sensor can be set below 0.1%/K in a Wheatstone bridge configuration with a direct current (dc) bias instead of a direct voltage bias (Cubells-Beltrán et al. 2011). Experimental parameters are only related to the nature of the GMR structures, and they have been measured elsewhere. In Figure 2.6, we show the typical values for full bridge sensors composed of equal SV elements, as

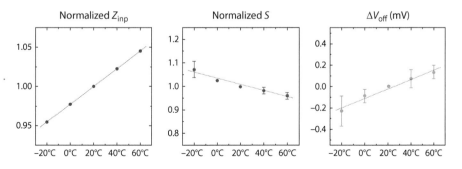

FIGURE 2.6
Temperature dependence of the characteristic parameters in typical spin valves. (From Cubells-Beltrán, M. et al., *IEEE Sensors Journal*, 9(12), 1756–62, 2009.)

described in Cubells-Beltrán et al. (2009). From these graphs, we can extract TCR ≈ 0.11%/°C, TCV$_{off}$ < 10 μV/°C, and TCS ≈ −0.15%/°C, as they were defined before.

When thermal drifts are not sufficiently lowered by using bridge configurations, different temperature compensation schemes have been proposed in the literature, specifically developed for GMR sensors–based applications (Ramírez Muñoz et al. 2006; Sánchez Moreno et al. 2011).

2.4.5 Bandwidth

Theoretically, due to the inherent quantum mechanism involved, MR structures have bandwidths of the order of 1 GHz (Hirota et al. 2002). In real applications, a reduction in the bandwidth is introduced by the associated circuitry. Such an effect can be important in electric current monitoring applications because of the necessity to drive the current path close to the sensor. In principle, due to the inductive character of the coupling, a "zero" behavior in the transfer function should be observed. In Figure 2.7, the frequency responses of some illustrative examples are compiled: an HMC1O21 sensor soldered onto a typical (3 mm width) printed circuit board (PCB) strap, a meandered engineered SV sensor (Reig et al. 2004), and an MTJ compact current sensor prototype both in a full Wheatstone bridge configuration and a single resistor (Reig et al. 2008). As observed, an inductive effect appears well below 1 MHz. Moreover, the more complicated the sensing structure is, the less bandwidth obtained. Regarding real applications, MR current sensors have been successfully applied up to 1 MHz (Cubells-Beltrán et al. 2009; Singh and Khambadkone 2014).

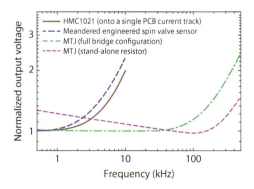

FIGURE 2.7
Frequency dependence of some selected GMR/MTJ current sensors. (From Cubells-Beltrán, M.-D. et al., *International Review of Electrical Engineering (IREE)* 6(1), 423–29, 2011.)

2.5 Applications

Now that we have described the principal characteristics of GMR/TMR sensors, we will describe some real applications in which these sensors have demonstrated their intrinsic capabilities.

2.5.1 General Purpose Magnetometers (Compass)

The measurement of the Earth's magnetic field is the ancient application of magnetic field sensors. The Earth's magnetic field ranges between approximately 25 and 60 µT, which is well covered by GMR/TMR sensors. With the increasing demand for digital compasses for mobile applications (mainly smartphones), GMR/TMR sensors have entered the market in serious competition with standard Hall solutions. Table 2.1 shows the main parameters of some selected MR compasses, including electronics (Reig et al. 2013).

2.5.2 Industrial Applications

2.5.2.1 Automotive

GMR sensors have entered the automotive market in several applications: steering angle measurement, rotor position measurement, speed sensing or crank shaft speed, and positions sensing. All of these can be classified as speed measurement or angle measurement. SV structures are commonly preferred for both sets of applications. A typical GMR angle sensor consists of several GMR resistors arranged in two bridges, one for each orthogonal direction (see Figure 2.8, left). Meandered geometries are used for increasing the total resistance to the kiloohms (kΩ) range. In this way, they provide a sine and a cosine signal that can be used in the calculation of the absolute angle of the magnetic field vector (see Figure 2.8, right). Due to this measurement principle, only the field direction, not the field magnitude, is relevant. The use of the GMR principle allows the measurement of angles in the full range of 360° in contrast to AMR-based sensors that cover only 180°. In any case, calibration of the devices is commonly required.

2.5.2.2 Space

In space sector applications, mass, volume, and power savings are important issues. GMR sensors are excellent candidates not only in planetary magnetometry, but also as magnetic encoders and angular or position sensors. It must be mentioned that space is an environment of extreme parameters, including wide temperature swings, very low pressures, moderate to high radiation, mechanical vibrations and impacts, and so on. GMR sensors have been used on several occasions in different satellite missions. As a summary, the evolution of MR

TABLE 2.1
Detailed Parameters of Selected Commercial Magnetoresistance-Based Compasses

General	Technology	AMR	AMR	GMR	MTJ
	Company	MEMSIC	Honeywell	Yamaha	Freescale
	Product	MMC314XMR	HMC5883L	YAS529	Mag3110
PKG	PKG	LGA10	LGA16	WLCSP10	DFN10
	Size (mm³)	3 × 3 × 1	3 × 3 × 0.9	2 × 2 × 1	2 × 2 × 0.85
I/O	Voltage (V)	1.7–3.6	2.7–5.25	2.16–3.6	1.95–3.6
	Working current (mA)	~2	~2	4	>1
	Samples per s @mA	50 @ 0.55	7.5 @ 0.10	4	10 @ 0.14
	Interface	I2C	I2C	I2C	I2C
	Interrupt	—	—	—	Y
Maximum ratings	Storage temp (°C)	−55/+125	−40/+125	−50/+125	−40/+125
	Operating temp (°C)	−40/+85	−30/+85	−40/+95	−40/+85
	Max exposed field (G)	—	10k	2k	1k
Performance	Range (+/− G)	4	1–8	3	10
	ADC (output bits)	12	12	10	15
	Resolution (mG)	2	2	6 (x,y)/12 (z)	1
	Offset (+/− G)	0.2	—	—	0.01
	Accuracy (°)	2	2	5	—
	Linearity (%FS)	1	0.1	—	1
	Hysteresis (%FS)	0.1	0.0025	—	1
	Repeatability (%FS)	—	0.1	—	—
	Sensitivity TC (%/°)	0.11	—	—	0.1
	Offset TC (mG/°)	0.4	—	—	0.1
	Bandwidth (Hz)	40	75	40	40
	Noise (RMS)	0.6 mG @ 25 Hz	—	—	0.5 mG
Features	On-chip temp sensor	Y	—	Y	Y
	Single-chip integration	—	—	Y	—
	Offset removal	Y	Y	—	—
	Self test	—	Y	—	—
	Other			3 external AD	Oversampling

Giant and Tunnel MR Sensors

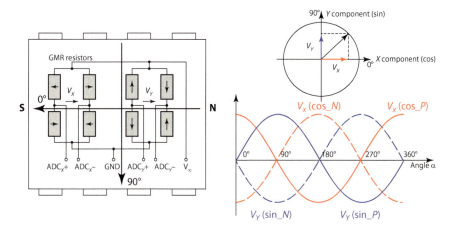

FIGURE 2.8
Typical arrangement of GMR angle sensors for automotive applications.

technologies in space applications is depicted in Figure 2.9 (Díaz-Michelena 2009).

2.5.2.3 Electric Current Measurement

Traditional methods for measuring an electric current include shunt resistors and the transformer principle, and its indirect measurement is by means of the generated magnetic field (Ziegler et al. 2009). This latter scheme has successfully approached making use of different magnetic field sensors (Pavel Ripka and Janosek 2010). In this sense, GMR sensors display some intrinsic properties, making them optimal for electric current measurement schemes, namely, high sensitivity, capable of integrating with other technologies (PCB or CMOS), and the measurement of in-plane magnetic fields. An extensive review on the application of GMR sensors in electrical current measurement can be found in Reig et al. (2009). The basic scheme is very simple: we just need to drive the desired current path to the neighborhood of the sensor, as depicted in Figure 2.10a. The sensor can be placed either above or below the current path. As observed, the magnetic field lines fall almost parallel to the sensor layers.

Sensing performance can be improved by using a bridge configuration. When the current path is already defined (usually a straight line), a scheme, as shown in Figure 2.10b1, with a half bridge is applicable. In this case, R_1 and R_3 are active and R_2 and R_4 are usually shielded as suggested, for example, in Singh and Khambadkone (2008) or Vieth et al. (2000). To get the full bridge behavior, we can fabricate the sensor with four active (opposite) elements (Pelegri Sebastia et al. 2004) or actuate on the design of the current path, as illustrated in Figure 2.10b2–b4. The successful application of these schemes can be found in Sanchez et al. (2012) and Pannetier-Lecoeur

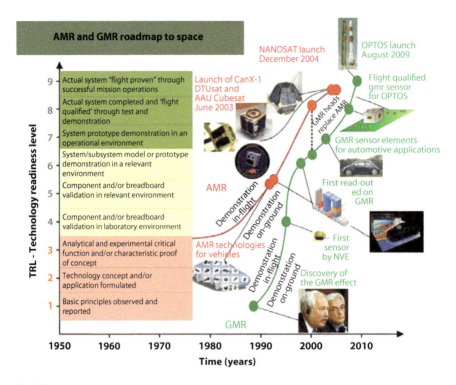

FIGURE 2.9
Recent history of AMR and GMR technologies in space applications. (From Díaz-Michelena, M., *Sensors (Basel, Switzerland)*, 9(4), 2271–88, 2009.)

et al. (2007) (Figure 2.10b2); Reig et al. (2004) and Cubells-Beltrán et al. (2009) (Figure 2.10b3); and Reig et al. 2005 and Cubells-Beltrán et al. (2009) (Figure 2.10b4). For an optimal geometric design of the current paths and the position of the sensors, numerical models (mainly the finite element method [FEM]) are commonly used (Beltran et al. 2007).

Limitations on the performance of GMR-based current sensors arise mainly from packaging issues (Cubells-Beltrán et al. 2011). Most significant is heating from the joule effect because of current and bandwidth constraints due to the capacitive/inductive coupling of the current path. Thermal drifts due to joule heating (Vopalensky and Platil 2013) can be significant not only in medium/high current applications, but also in integrated circuits (ICs) environments. They can be reduced either with the use of full Wheatstone bridge sensors (Reig et al. 2004; Cubells-Beltrán et al. 2009) or with the use of external compensation electronic circuitry (Ramírez Muñoz et al. 2006; Sánchez Moreno et al. 2011). To properly analyze such effects, numerical modeling is commonly used, from both a physical (Beltran et al. 2007) and an electrical point of view (Roldán et al. 2010a).

Giant and Tunnel MR Sensors 53

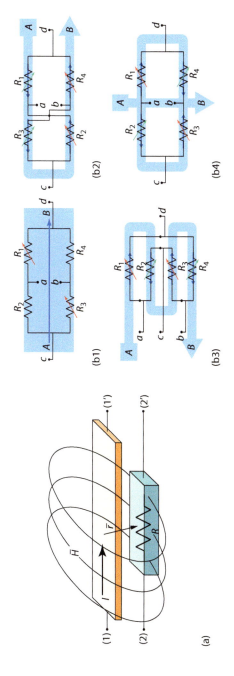

FIGURE 2.10
Current measurement with GMR sensors: (a) principle description, (b) different geometrical approaches.

Specific applications of GMR sensors for electrical current measurement include built-in current sensing (BICS) schemes in ICs. This was first demonstrated in 2005 (Reig et al. 2005) and then improved with low noise devices (Pannetier-Lecoeur et al. 2007), using MTJ sensors (Le Phan et al. 2005), full bridge arrangements (Cubells-Beltrán et al. 2009), and improved conditioning circuitry (Madrenas et al. 2014). Electric currents in the range of 1 µA have been resolved in this way. In addition, these schemes have become successful when integrated with standard CMOS technologies (Cubells-Beltrán et al. 2014).

We should also mention that GMR sensors have also been used in related applications such as analog electric isolators (Reig et al. 2008) and integrated wattmeters (Roldán et al. 2010b).

2.5.3 Non-Destructive Evaluation

Non-destructive evaluation (NDE) refers to any examination, test, or evaluation performed on any type of object without changing or altering it in any way, in order to determine the absence (or presence) of discontinuities that could jeopardize its functionality. The most common NDE methods are optical inspection (including non-visible range such as infrared or x-rays), ultrasonic tests, and magnetic measurements.

2.5.3.1 Magnetic Detection

As mentioned, magnetic measurements are commonly used in NDE in general and scan systems (imaging or detecting) in particular. We can measure the magnetic field of the scanned objects (if existing) or the perturbation that they produce in the Earth's magnetic field. Traditionally, pick-up coils and Hall sensors have been used. We will enumerate some of the successful applications in which GMR sensors have been considered in these scenarios.

GMR sensors have been successfully used for the detection/monitoring of different kinds of objects, including traffic speed monitoring (Pelegrí Sebastiá et al. 2007), tool vibration (Sebastia et al. 2009), weapon detection (Tian et al. 2012), localization of hidden metallic objects (Renhart et al. 2011), robot movement control (Ku et al. 2000), and even electric guitar monitoring (Lenssen et al. 2002).

Regarding specific imaging techniques (also scanning microscopy), GMR sensors have been applied to the evaluation of current faults at the IC level with sub-micron resolution involving electric currents below 1 mA (Reig et al. 2013). The magnetic field microscopy of rock samples using a GMR-based scanning magnetometer has also been reported (Hankard et al. 2009).

2.5.3.2 Eddy Current Testing

Within magnetic field–based techniques, eddy current testing (ECT) has been specifically considered in a wide range of modern testing processes, including defects in metallic surfaces and subsurfaces.

Eddy currents are closed loops of induced current circulating in a plane that is perpendicular to the direction of a time-varying magnetic flux density B, as depicted in Figure 2.11. The variation of B generates an electric field intensity E, in a loop as expressed by the Maxwell equation:

$$\nabla \times \mathbf{E} = -\frac{\partial \mathbf{B}}{\partial t}$$

Therefore, the current density J, in a material with conductivity σ, also circulates in the loop, because

$$\mathbf{J} = \sigma \mathbf{E}$$

Such currents, due to conductivity σ, can degrade the performance of the affected surfaces, but they can also be used for detecting flaws or cracks on metallic materials, conductivity variations, spacing between probe and device under test, material thickness, thickness of platting or cladding on a base material, spacing between conductive layers, or permeability variations.

The concept of using GMR in ECT measurements was first introduced by Dogaru and Smith (2001), through two geometrical approaches taking advantage of the inherent properties of GMR sensors, as detailed in Figure 2.11. In this particular experiment, currents of the order of 1 A at 30 kHz were used for the exciting coils, and NVE commercial unipolar devices (~25 mV/V/mT sensitivity, 2 mT, linear range) as GMR sensors. In this way, cracks 1–15 mm long, 0.5 mm wide, and 0.25–4 mm deep were

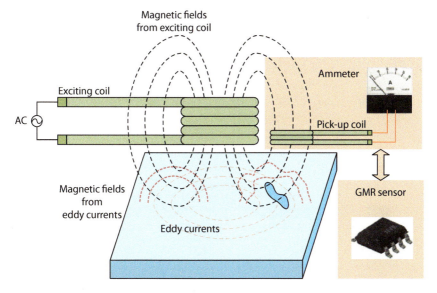

FIGURE 2.11
Scheme of an eddy current testing system, with a GMR as sensing element.

scanned. A similar system with portable characteristics is described in Betta et al. 2012). By including a feedback coil in the scheme in Figure 2.12, resolution in depth can be achieved (Jeng et al. 2006). The depth of the defects can also be detected (Espina-Hernández et al. 2012). In this scenario, the use of numerical models (mainly based on FEM) is highly helpful (Zeng et al. 2011).

ECT based on GMR sensors has been extensively applied to the evaluation of PCBs. Initially, Chomsuwan and co-workers (Yamada et al. 2004) demonstrated it with a specifically designed SV sensor (200 mV/mT sensitivity) with a printed meandered coil, by mapping PCB defects of the order of 100 µm size. These results were better than those obtained with a pick-up coil–based system. Novel experiments were performed on high-density double-layer PCBs with an improved ECT probe including an optimized meander coil and an array of GMR sensors (Chomsuwan et al. 2005, 2007a,b). Defects with sizes below 100 µm were resolved on both sides of the PCB. This topic is revisited frequently (Cacciola et al. 2011).

The ECT technique based on GMR sensors has also been applied to the evaluation of the health of an aircraft's structure (Nair et al. 2006). Pulsed currents have been demonstrated to improve the performance of ECT probes in this sense (Tamburrino et al., 2010).

2.5.4 Bio-Applications

Magnetic fields (generated and/or measured) are extensively used in biologic sciences including genetics, bio-technology, different fields of medicine (physiology, oncology, etc.), among others. Most of these applications require the measurement of very low magnetic fields (below the nT limit) in small spaces (commonly in the sub-mm range). At the beginning of the century, MR sensors started to be explored as the sensing elements in biochips. A biosensor can be defined as a "compact analytical device or unit incorporating a biological or biologically-derived sensitive element integrated or associated with a physio-chemical transducer" (Graham et al. 2004).

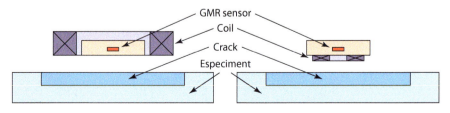

FIGURE 2.12
Different arrangements for integrating GMR sensors with exciting coils in ECT systems.

2.5.4.1 Detection of Bioanalytes

General bioanalytes (molecules, cells, viruses, bacteria, tissues) are not magnetic. In order to take advantage of GMR for monitoring or detecting bioanalytes, they must be bonded to magnetic elements (usually nanoparticles [NPs]) (Wang et al. 2014; Lee et al. 2009) and driven near the sensor by means of microfluidics (Muluneh and Issadore 2014) or guiding magnetic fields (Giouroudi and Keplinger 2013; Gooneratne et al. 2011). GMR sensors have been successfully applied to the detection of proteins (Gaster et al. 2011), DNA (Wang et al. 2013), viruses (Zhi et al. 2014), and bacteria (Mujika et al. 2009). In this way, two approaches can be defined: static and dynamic, as depicted in Figure 2.13.

An example of the static approach is the analysis of DNA (Koets et al. 2009). For DNA detection, single-stranded DNA receptors are first immobilized on the surface of magnetic sensors. Oligonucleotides of unknown sequence are selectively captured by complementary probes. SA-coated magnetic NPs are then introduced and bind to the biotin of the hybridized DNA. Finally, magnetic field disturbances because of the NPs are sensed by magnetic sensors. Biotin and streptavidin are often used in this process (Wang et al. 2013). To improve the performance of the sensor system, microcoils can be integrated in association with the sensing elements. These coils generate a magnetic field that is used to attract the magnetic beads to the sensor area and activate them (Freitas et al. 2011). In this way, the femtomolar limit of detection has been achieved.

For the detection of general cells (cytometry), magnetic nanobeads need to be bonded to them (Freitas et al. 2012). Then, by means of microfluidics or guiding magnetic fields, they are driven close to the sensors, where the detection is performed, as described in Shoshi et al. (2012).

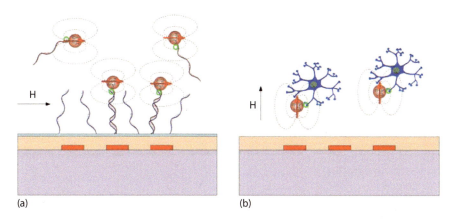

FIGURE 2.13
Static (a) and dynamic (b) approaches for bioanalytes detection with GMR sensors.

2.5.4.2 Monitoring of Magnetic Fluids

Fluids incorporating magnetic particles (usually nanobeads) are known as *magnetic fluids*. They can be made biocompatible for *in vivo* applications, such as hyperthermia cancer therapy. A magnetic fluid is injected into the affected area and an external ac magnetic flux density is applied to exploit the self-heating properties of the magnetic beads in the fluid. Temperatures in excess of 42°C destroy tumors (Reig et al. 2013). Hence, the accurate estimation of magnetic fluid content density is critical for the success of the treatment. In Mukhopadhyay et al. (2007), a GMR-based needle probe 20 mm long and 310 µm wide, comprising four SV sensors was analyzed. The needle probe was successfully tested in tumor-simulating cylindrical agar cavities.

2.5.4.3 Biomedical Signal Detection

GMR/TMR-based micromachined probe needles have also been designed for detecting brain activity through the measurement of generated magnetic fields. After demonstrating the concept (Amaral et al. 2011), MTJ-based microneedles were developed and characterized (Amaral et al. 2013). The associated electronics were also specifically implemented (Costa et al. 2014).

A non-invasive detection system for magnetocardiography applications has also been described by Pannetier-Lecoeur et al. (2010), where a GMR sensor is integrated with a high-temperature superconductor (YBCO).

2.6 Conclusions

GMR/TMR sensors have rapidly passed their initial potentiality in real applications and have become the first option in a huge number of scenarios demanding the measurement of low magnetic fields with a high level of integration devices. Complete knowledge of the underlying phenomena has allowed the specific design of different kinds of devices. Their SNR, and subsequent detectivity, make them suitable for most applications in different fields such as bio-technology, microelectronics, and space.

Acknowledgments

The authors are grateful for the very fruitful collaborations that have made possible some of the results presented in this chapter: INESC-MN (Lisboa, Portugal), University of L'Aquila (Italy), UPC (Barcelona, Spain), and CNM (Barcelona, Spain).

References

Ač, V. 2008. A study of hysteresis in the GMR layer structures by FEM. *Physica B: Condensed Matter* 403 (2–3): 460–63.

Amaral, J., S. Cardoso, P. P. Freitas, and A. M. Sebastião. 2011. Toward a system to measure action potential on mice brain slices with local magnetoresistive probes. *Journal of Applied Physics* 109 (7): 07B308.

Amaral, J., V. Pinto, T. Costa, J. Gaspar, R. Ferreira, E. Paz, S. Cardoso, and P. P. Freitas. 2013. Integration of TMR sensors in silicon microneedles for magnetic measurements of neurons. *IEEE Transactions on Magnetics* 49 (7): 3512–15.

Andrés, J. P., J. Colino, and J. M. Riveiro. 1999. Enhancement of GMR in as-deposited Co-Cu granular films with RF sputtering power. *Journal of Magnetism and Magnetic Materials* 196–197 (May): 493–94.

Baibich, M. N., J. M. Broto, A. Fert, F. N. Vandau, F. Petroff, P. Eitenne, G. Creuzet, A. Friederich, and J. Chazelas. 1988. Giant magnetoresistance of (001)Fe/(001)Cr magnetic superlattices. *Physical Review Letters* 61 (21): 2472–75.

Beltran, H., C. Reig, V. Fuster, D. Ramírez, and M. D. Cubells-Beltrán. 2007. Modeling of magnetoresistive-based electrical current sensors: A technological approach. *IEEE Sensors Journal* 7 (11): 1532–37.

Betta, G., L. Ferrigno, and M. Laracca. 2012. GMR-based ECT instrument for detection and characterization of crack on a planar specimen: A hand-held solution. *IEEE Transactions on Instrumentation and Measurement* 61 (2): 505–12.

Binasch, G., P. Grunberg, F. Saurenbach, and W. Zinn. 1989. Enhanced magnetoresistance in layered magnetic-structures with antiferromagnetic interlayer exchange. *Physical Review B* 39 (7): 4828–30.

Cacciola, M., G. Megali, D. Pellicanó, and F. C. Morabito. 2011. A GMR–ECT based embedded solution for applications on PCB inspections. *Sensors and Actuators A: Physical* 167 (1): 25–33.

Cardoso, F. A., H. A. Ferreira, J. P. Conde, V. Chu, P. P. Freitas, D. Vidal, J. Germano, et al. 2006. Diode/magnetic tunnel junction cell for fully scalable matrix-based biochip. *Journal of Applied Physics* 99 (8): 08B307.

Chen, X., and P. P. Freitas. 2012. Magnetic tunnel junction based on MgO barrier prepared by natural oxidation and direct sputtering deposition. *Nano-Micro Letters* 4 (1): 25–29.

Chomsuwan, K., S. Yamada, and M. Iwahara. 2007a. Improvement on defect detection performance of PCB inspection based on ECT technique with multi-SV-GMR sensor. *IEEE Transactions on Magnetics* 43 (6): 2394–96.

Chomsuwan, K., S. Yamada, and M. Iwahara. 2007b. Bare PCB inspection system with SV-GMR sensor eddy-current testing probe. *IEEE Sensors Journal* 7 (5): 890–96.

Chomsuwan, K., S. Yamada, M. Iwahara, H. Wakiwaka, and S. Shoji. 2005. Application of eddy-current testing technique for high-density double-layer printed circuit board inspection. *IEEE Transactions on Magnetics* 41 (10): 3619–21.

Costa, T., M. S. Piedade, J. Germano, J. Amaral, and P. P. Freitas. 2014. A neuronal signal detector for biologically generated magnetic fields. *IEEE Transactions on Instrumentation and Measurement* 63 (5): 1171–80.

Cubells-Beltrán, M. D., C. Reig, A. De Marcellis, E. Figueras, A. Yúfera, B. Zadov, E. Paperno, S. Cardoso, and P. P. Freitas. 2014. Monolithic integration of giant magnetoresistance (GMR) devices onto standard processed CMOS dies. *Microelectronics Journal* 45 (6): 702–7.

Cubells-Beltrán, M.-D., C. Reig, J. Martos, J. Torres, and J. Soret. 2011. Limitations of magnetoresistive current sensors in industrial electronics applications. *International Review of Electrical Engineering (IREE)* 6 (1): 423–29.

Cubells-Beltrán, M. D., C. Reig, D. R. Muñoz, S. I. P. C. de Freitas, and P. J. P. De Freitas. 2009. Full Wheatstone bridge spin-valve based sensors for IC currents monitoring. *IEEE Sensors Journal* 9 (12): 1756–62.

Díaz-Michelena, M. 2009. Small magnetic sensors for space applications. *Sensors (Basel, Switzerland)* 9 (4): 2271–88.

Dogaru, T., and S. T. Smith. 2001. Giant magnetoresistance-based eddy-current sensor. *IEEE Transactions on Magnetics* 37 (5): 3831–38.

Espina-Hernández, J. H., E. Ramírez-Pacheco, F. Caleyo, J. A. Pérez-Benitez, and J. M. Hallen. 2012. Rapid estimation of artificial near-side crack dimensions in aluminium using a GMR-based eddy current sensor. *NDT & E International* 51 (October): 94–100.

Ferreira, R., P. Wisniowski, P. P. Freitas, J. Langer, B. Ocker, and W. Maass. 2006. Tuning of MgO barrier magnetic tunnel junction bias current for picotesla magnetic field detection. *Journal of Applied Physics* 99 (8): 08K706.

Freitas, P. P., F. A. Cardoso, V. C. Martins, S. A. M. Martins, J. Loureiro, J. Amaral, R. C. Chaves, et al. 2012. Spintronic platforms for biomedical applications. *Lab on a Chip* 12 (3): 546–57.

Freitas, P. P., S. Cardoso, R. Ferreira, V. C. Martins, A. Guedes, F. A. Cardoso, J. Loureiro, R. Macedo, R. C. Chaves, and J. Amaral. 2011. Optimization and integration of magnetoresistive sensors. *SPIN* 01 (01): 71–91.

Freitas, P. P., R. Ferreira, S. Cardoso, and F. Cardoso. 2007. Magnetoresistive sensors. *Journal of Physics: Condensed Matter* 19 (16). IOP Publishing: 165221.

Gaster, R. S, L. Xu, S-J. Han, R. J. Wilson, D. A. Hall, S. J. Osterfeld, H. Yu, and S. X. Wang. 2011. Quantification of protein interactions and solution transport using high-density GMR sensor arrays. *Nature Nanotechnology* 6 (5): 314–20.

Giouroudi, I., and F. Keplinger. 2013. Microfluidic biosensing systems using magnetic nanoparticles. *International Journal of Molecular Sciences* 14 (9): 18535–56.

Gooneratne, C. P., C. Liang, and J. Kosel. 2011. A planar conducting microstructure to guide and confine magnetic beads to a sensing zone. *Microelectronic Engineering* 88 (8): 1757–60.

Graham, D. L., H. A. Ferreira, and P. P. Freitas. 2004. Magnetoresistive-based biosensors and biochips. *Trends in Biotechnology* 22 (9): 455–62.

Guerrero, R., M. Pannetier-Lecoeur, C. Fermon, S. Cardoso, R. Ferreira, and P. P. Freitas. 2009. Low frequency noise in arrays of magnetic tunnel junctions connected in series and parallel. *Journal of Applied Physics* 105 (11): 113922.

Hall, D. A., R. S. Gaster, K. Makinwa, S. X. Wang, and B. Murmann. 2013. A 256 pixel magnetoresistive biosensor microarray in 0.18μm CMOS. *IEEE Journal of Solid-State Circuits* 48 (5): 1290–1301.

Han, J., J. Hu, Y. Ouyang, S. X. Wang, and J. He. 2015. Hysteretic modeling of output characteristics of giant magnetoresistive current sensors. *IEEE Transactions on Industrial Electronics* 62 (1): 516–24.

Hankard, F., J. Gattacceca, C. Fermon, M. Pannetier-Lecoeur, B. Langlais, Y. Quesnel, P. Rochette, and S. A. McEnroe. 2009. Magnetic field microscopy of rock samples using a giant magnetoresistance-based scanning magnetometer. *Geochemistry, Geophysics, Geosystems* 10: Q10Y06.

He, H., K. Zhernenkov, M. Vadalá, N. Akdogan, D. Gorkov, R. M. Abrudan, B. P. Toperverg, H. Zabel, H. Kubota, and S. Yuasa. 2010. The effect of annealing on the junction profile of CoFeB/MgO tunnel junctions. *Journal of Applied Physics* 108 (6): 063922.
Hirota, E., H. Sakakima, and K. Inomata. 2002. *Giant Magneto-Resistance Devices*. Vol. 40. Springer Series in Surface Sciences. Berlin: Springer. doi:10.1007/978-3-662-04777-4.
Hooge, F. N. 1976. $1/f$ noise. *Physica B+C* 83 (1): 14–23.
Jedlicska, I., R. Weiss, and R. Weigel. 2010. Linearizing the output characteristic of GMR current sensors through hysteresis modeling. *IEEE Transactions on Industrial Electronics* 57 (5): 1728–34.
Jeng, J.-T., G-S. Lee, W-C. Liao, and C-L. Shu. 2006. Depth-resolved eddy-current detection with GMR magnetometer. *Journal of Magnetism and Magnetic Materials* 304 (1): e470–73.
Johnson, J. B. 1928. Thermal agitation of electricity in conductors. *Physical Review* 32 (1): 97–109.
Koets, M., T. van der Wijk, J. T. W. M. van Eemeren, A. van Amerongen, and M. W. J. Prins. 2009. Rapid DNA multi-analyte immunoassay on a magneto-resistance biosensor. *Biosensors and Bioelectronics* 24 (7): 1893–98.
Ku, W., P. P. Freitas, P. Compadrinho, and J. Barata. 2000. Precision X-Y robotic object handling using a dual GMR bridge sensor. *IEEE Transactions on Magnetics* 36 (5): 2782–84.
Le, P., K., H. Boeve, F. Vanhelmont, T. Ikkink, and W. Talen. 2005. Geometry optimization of TMR current sensors for on-chip IC testing. *IEEE Transactions on Magnetics* 41 (10): 3685–87.
Lee, K., S. Lee, B. K. Cho, K-S. Kim, and B. Kim. 2009. The limit of detection of giant magnetoresistive (GMR) sensors for bio-applications. *Journal of the Korean Physical Society* 55 (1): 193–96.
Lenssen, K.-M. H., G. H. J. Somers, and J. B. A. D. van Zon. 2002. Magnetoresistive sensors for string instruments. *Journal of Applied Physics* 91 (10): 7777.
Liu, S., Q. Huang, Y. Li, and W. Zhen. 2012. Experimental research on hysteresis effects in GMR sensors for analog measurement applications. *Sensors and Actuators A: Physical* 182 (August): 72–81.
Lopez-Martin, A. J., and A. Carlosena. 2009. Performance tradeoffs of three novel GMR contactless angle detectors. *IEEE Sensors Journal* 9 (3): 191–98.
Madrenas, J., M.-D. Cubells-Beltrán, B. Zadov, S. Cardoso, C. Reig, P. P. Freitas, E. Paperno, and A. De Marcellis. 2014. Quasi-digital front-ends for current measurement in integrated circuits with giant magnetoresistance technology. *IET Circuits, Devices & Systems* 8 (4): 291–300.
Mujika, M., S. Arana, E. Castaño, M. Tijero, R. Vilares, J. M. Ruano-López, A. Cruz, L. Sainz, and J. Berganza. 2009. Magnetoresistive immunosensor for the detection of *Escherichia coli* O157:H7 including a microfluidic network. *Biosensors & Bioelectronics* 24 (5): 1253–58.
Mukhopadhyay, S. C., K. Chomsuwan, C. P. Gooneratne, and S. Yamada. 2007. A novel needle-type SV-GMR sensor for biomedical applications. *IEEE Sensors Journal* 7 (3): 401–8.
Muluneh, M., and D. Issadore. 2014. A multi-scale PDMS fabrication strategy to bridge the size mismatch between integrated circuits and microfluidics. *Lab on a Chip* 14 (23): 4552–58.

Nair, N. V., V. R. Melapudi, H. R. Jimenez, X. Liu, Y. Deng, Z. Zeng, L. Udpa, T. J. Moran, and S. S. Udpa. 2006. A GMR-based eddy current system for NDE of aircraft structures. *IEEE Transactions on Magnetics* 42 (10): 3312–14.
Nyquist, H. 1928. Thermal agitation of electric charge in conductors. *Physical Review* 32 (1): 110–13.
Pannetier-Lecoeur, M., C. Fermon, A. de Vismes, E. Kerr, and L. Vieux-Rochaz. 2007. Low noise magnetoresistive sensors for current measurement and compasses. *Journal of Magnetism and Magnetic Materials* 316 (2): e246–48.
Pannetier-Lecoeur, M., C. Fermon, H. Dyvorne, J. F. Jacquinot, H. Polovy, and A. L. Walliang. 2010. Magnetoresistive-superconducting mixed sensors for biomagnetic applications. *Journal of Magnetism and Magnetic Materials* 322 (9–12): 1647–50.
Pelegrí Sebastiá, J., J. Alberola Lluch, and J. Rafael Lajara Vizcaíno. 2007. Signal conditioning for GMR magnetic sensors applied to traffic speed monitoring. *Sensors and Actuators A: Physical* 137 (2): 230–35.
Pelegri Sebastia, J., D. R. Munoz, and P. J. P. de Freitas. 2004. A novel spin-valve bridge sensor for current sensing. *IEEE Transactions on Instrumentation and Measurement* 53 (3): 877–80.
Peña, V., Z. Sefrioui, D. Arias, C. Leon, J. Santamaria, J. L. Martinez, S. G. E. Te Velthuis, and A. Hoffmann. 2005. Giant magnetoresistance in ferromagnet/superconductor superlattices. *Physical Review Letters* 94 (5): 057002.
Pullini, D., D. Busquets, A. Ruotolo, G. Innocenti, and V. Amigó. 2007. Insights into pulsed electrodeposition of GMR multilayered nanowires. *Journal of Magnetism and Magnetic Materials* 316 (2): e242–45.
Ramírez Muñoz, D., J. Sánchez Moreno, S. Casans Berga, E. Castro Montero, C. Reig Escrivà, and A. Edith Navarro Antón. 2006. Temperature compensation of Wheatstone bridge magnetoresistive sensors based on generalized impedance converter with input reference current. *Review of Scientific Instruments* 77 (10): 105102.
Ranchal, R., M. Torija, E. López, M. C. Sánchez, C. Aroca, and P. Sánchez. 2002. The influence of anisotropy on the magnetoresistance of permalloy-copper-permalloy thin films. *Nanotechnology* 13 (3): 392–97.
Reig, C., S. Cardoso, and S. Mukhopadhyay. 2013. *Giant Magnetoresistance (GMR) Sensors: From Basis to State-of-the-Art Applications*. Berlin: Springer.
Reig, C., M.-D. Cubells-Beltran, D. Ramirez, S. Cardoso, and P.P. Freitas. 2008. Electrical isolators based on tunneling magnetoresistance technology. *IEEE Transactions on Magnetics* 44 (11): 4011–14.
Reig, C., M.-D. Cubells-Beltran, and D. Ramírez Muñoz. 2009. Magnetic field sensors based on giant magnetoresistance (GMR) technology: Applications in electrical current sensing. *Sensors (Basel, Switzerland)* 9 (10): 7919–42.
Reig, C., D. Ramirez, H. H. Li, and P. P. Freitas. 2005. Low-current sensing with specular spin valve structures. *IEE Proceedings: Circuits, Devices and Systems* 152 (4): 307–11.
Reig, C., D. Ramírez, F. Silva, J. Bernardo, P. Freitas, D. Ramírez, F. Silva, J. Bernardo, and P. Freitas. 2004. Design, fabrication, and analysis of a spin-valve based current sensor. *Sensors and Actuators, A: Physical* 115: 259–66.
Renhart, W., M. Bellina, C. Magele, and A. Köstinger. 2011. Hidden metallic object localization by using giant magnetic resistor sensors. *COMPEL: The International Journal for Computation and Mathematics in Electrical and Electronic Engineering* 30 (6): 1927–37.

Ripka, P., and M. Janosek. 2010. Advances in magnetic field sensors. *IEEE Sensors Journal* 10 (6): 1108–16.
Ripka, P., M. Tondra, J. Stokes, and R. Beech. 1999. AC-driven AMR and GMR magnetoresistors. *Sensors and Actuators A: Physical* 76 (1–3): 225–30.
Roldán, A., C. Reig, M. D. Cubells-Beltrán, J. B. Roldán, D. Ramírez, S. Cardoso, and P. P. Freitas. 2010a. Analytical compact modeling of GMR based current sensors: Application to power measurement at the IC level. *Solid-State Electronics* 54 (12): 1606–12.
Roldán, A., C. Reig, M. D. D. Cubells-Beltrán, J. B. B. Roldán, D. Ramírez, S. Cardoso, and P. P. P. Freitas. 2010b. Analytical compact modeling of GMR based current sensors: Application to power measurement at the IC level. *Solid-State Electronics* 54 (12): 1606–12.
Sanchez, J., D. Ramirez, S. I. Ravelo, A. Lopes, S. Cardoso, R. Ferreira, and P. P. Freitas. 2012. Electrical characterization of a magnetic tunnel junction current sensor for industrial applications. *IEEE Transactions on Magnetics* 48 (11): 2823–26.
Sánchez Moreno, J., D. Ramírez Muñoz, S. Cardoso, S. Casans Berga, A. E. Navarro Antón, and P. J. Peixeiro de Freitas. 2011. A non-invasive thermal drift compensation technique applied to a spin-valve magnetoresistive current sensor. *Sensors (Basel, Switzerland)* 11 (3): 2447–58.
Sebastia, J. P., J. Alberola Lluch, J. R. Lajara Vizcaino, and J. Santiso Bellon. 2009. Vibration detector based on GMR sensors. *IEEE Transactions on Instrumentation and Measurement* 58 (3): 707–12.
Shoshi, A., J. Schotter, P. Schroeder, M. Milnera, P. Ertl, V. Charwat, M. Purtscher, et al. 2012. Magnetoresistive-based real-time cell phagocytosis monitoring. *Biosensors & Bioelectronics* 36 (1): 116–22.
Singh, R. P., and A. M. Khambadkone. 2008. Giant magneto resistive (GMR) effect based current sensing technique for low voltage/high current voltage regulator modules. *IEEE Transactions on Power Electronics* 23 (2): 915–25.
Singh, R. P., and A. M. Khambadkone. 2014. A giant magneto resistive (GMR) effect based current sensor with a toroidal magnetic core as flux concentrator and closed-loop configuration. *IEEE Transactions on Applied Superconductivity* 24 (3): 1–5.
Stutzke, N. A., S. E. Russek, D. P. Pappas, and M. Tondra. 2005. Low-frequency noise measurements on commercial magnetoresistive magnetic field sensors. *Journal of Applied Physics* 97 (10): 10Q107.
Tamburrino, A., L. Udpa, and S. S. Udpa. 2010. Pulsed eddy-current based giant magnetoresistive system for the inspection of aircraft structures. *IEEE Transactions on Magnetics* 46 (3): 910–17.
Thompson, S. M. 2008. The discovery, development and future of GMR: The Nobel Prize 2007. *Journal of Physics D: Applied Physics* 41 (9): 093001.
Tian, G. Y., A. Al-Qubaa, and J. Wilson. 2012. Design of an electromagnetic imaging system for weapon detection based on GMR sensor arrays. *Sensors and Actuators A: Physical* 174 (February): 75–84.
Vieth, M., W. Clemens, H. van den Berg, G. Rupp, J. Wecker, and M. Kroeker. 2000. Contactless current detection with GMR sensors based on an artificial antiferromagnet (AAF) subsystem. *Sensors and Actuators A: Physical* 81 (1–3): 44–48.
Vopalensky, M., and A. Platil. 2013. Temperature drift of offset and sensitivity in full-bridge magnetoresistive sensors. *IEEE Transactions on Magnetics* 49 (1): 136–39.

Vopálenský, M., P. Ripka, J. Kubík, and M. Tondra. 2004. Improved GMR sensor biasing design. *Sensors and Actuators A: Physical* 110 (1–3): 254–58.

Wang, W., Y. Wang, L. Tu, Y. Feng, T. Klein, and J.-P. Wang. 2014. Magnetoresistive performance and comparison of supermagnetic nanoparticles on giant magnetoresistive sensor-based detection system. *Scientific Reports* 4 (January): 5716.

Wang, W., Y. Wang, L. Tu, T. Klein, Y. Feng, and J.-P. Wang. 2013. Surface modification for protein and DNA immobilization onto GMR biosensor. *IEEE Transactions on Magnetics* 49 (1): 296–99.

Yamada, S., K. Chomsuwan, Y. Fukuda, M. Iwahara, H. Wakiwaka, and S. Shoji. 2004. Eddy-current testing probe with spin-valve type GMR sensor for printed circuit board inspection. *IEEE Transactions on Magnetics* 40 (4): 2676–78.

Zeng, Z., Y. Deng, X. Liu, L. Udpa, S. S. Udpa, B. E. C. Koltenbah, R. H. Bossi, and G. Steffes. 2011. EC-GMR data analysis for inspection of multilayer airframe structures. *IEEE Transactions on Magnetics* 47 (12): 4745–52.

Zhi, X., M. Deng, H. Yang, G. Gao, K. Wang, H. Fu, Y. Zhang, D. Chen, and D. Cui. 2014. A novel HBV genotypes detecting system combined with microfluidic chip, loop-mediated isothermal amplification and GMR sensors. *Biosensors & Bioelectronics* 54 (April): 372–77.

Ziegler, S., R. C. Woodward, H. Ho-Ching Iu, and L. J. Borle. 2009. Current sensing techniques: A review. *IEEE Sensors Journal* 9 (4): 354–76.

Ziese, M., and M. J. Thornton. 2001. *Spin Electronics. Lecture Notes in Physics*. Berlin: Springer-Verlag.

3

Frequency Tuning Investigation of an Out-of-Plane Resonant Microstructure for a Capacitive Detection Magnetometer

Petros Gkotsis, Mohamed Hadj Said, Farès Tounsi, Brahim Mezghani, and Laurent A. Francis

CONTENTS

3.1 Introduction 66
3.2 Lorentz Force–Based Capacitive Magnetometers 67
    3.2.1 In-Plane Capacitive Detection 67
    3.2.2 Out-of-Plane Capacitive Detection 70
3.3 Magnetometer with Capacitive Detection 74
3.4 Mechanical and Electrical Behavior 76
    3.4.1 Magnetometer Mechanical Model Analysis 76
    3.4.2 Electric Model Analysis 79
    3.4.3 Magnetometer Sensitivity Evaluation 81
3.5 Frequency Tuning by Applying dc Voltage 85
    3.5.1 Physical Modeling 86
    3.5.2 Numerical Resolution 89
    3.5.3 Experimental Validation and Discussion 95
3.6 Conclusion 98
References 98

ABSTRACT In this chapter, we are interested in resonant magnetic sensors using capacitive detection. Specifically, we will analyze their sensing behavior and their performances. The device under test, the xylophone magnetometer, is a sensor designed and fabricated at the Université Catholique de Louvain (UCL), Belgium. This design is different from other designs generally reported in literature because it relies on interdigitated finger electrodes for sensing. Moreover, we will study potential problems that can provide an erroneous operation of this kind of magnetometer design and report possible solutions.

3.1 Introduction

Micromechanical resonators are the basic components of numerous microdevices, such as accelerometers [1,2], microelectromechanical filters [3–5], gyroscopes [6,7], and magnetometers [8,9]. Resonant magnetometers are magnetic field sensors that exploit the Lorentz force exerted on resonating micromachined structures. It is a kind of inertial microelectromechanical system (MEMS) that uses a spring-mass resonant system. Indeed, when an alternating current (ac) passes through the resonant structure and is coupled to a magnetic field, a Lorentz force is generated and the resulting displacement of the structure can be measured either by optical, piezoresistive, piezoelectric, or capacitive sensing techniques [10]. If the frequency of the ac is set equal to one of the resonance frequencies of the suspended structure (generally chosen the first), the amplitude of the resulting vibration will be maximum. These resonant sensors can detect magnetic fields up to 1 T with a resolution down to 1 nT, and have a wide sensitivity range [11]. They could compete with fluxgate sensors and Hall sensors in numerous applications where the measurement of magnetic fields is a requirement [11]. In this chapter, we are interested in resonant magnetic sensors using capacitive detection. Specifically, we will analyze their sensing behavior and their performances. The device under test, the xylophone magnetometer, is a sensor that has been designed and fabricated at the Université Catholique de Louvain (UCL), Belgium. This design is different from other designs generally reported in the literature because it relies on interdigitated finger electrodes for sensing. Moreover, we will study potential problems that can provide erroneous operation of this kind of magnetometer design and report possible solutions.

This chapter is organized as follows: in the first section, we discuss the operational principle of capacitive magnetometers for both in-plane and out-of-plane detection reported in the literature. We then present the xylophone bar magnetometer designed at the UCL and its operation, and the mechanical and electrical properties of this microsensor are evaluated. Next, the main part of this chapter, is a detailed method for tuning the resonant frequency of the microresonant clamped-clamped beam by applying an external potential. The resonant frequency of the structure is tuned by applying a direct current (dc) bias voltage between the interdigitated comb fingers defined on both the moving part of the sensor and the stationary part of the substrate, in order to control the spring constant of the moving part. It has been observed that an applied dc voltage causes the resonant frequency to shift to higher values, especially for small vibrations. A simple analytic model has been developed to accurately describe the shift. The measured resonance frequency of the clamped-clamped beam structure changed by up to 38% from its original value (around 18.8 kHz) when a bias voltage of 52 V was applied. Tuning the resonant frequency of the resonating structure has many advantages for the magnetometer as it can serve as a feedback mechanism for error compensation.

3.2 Lorentz Force–Based Capacitive Magnetometers

Capacitive detection is widely used in many applications due to its affordability and superior performance in terms of noise and low power consumption [12]. Specifically, the capacitive technique is commonly used in MEMS magnetometers [12]. Generally, the magnetometer comprises a resonant structure that vibrates due to the Lorentz force when a magnetic field interacts with a current passing through the resonant structure. This vibration causes a periodic capacitance variation between the fixed and the moving electrodes, which can be measured. This change in capacitance, which is generally part of an electronic resonating circuit, introduces a variation in the resonant frequency and/or in the output voltage. Therefore, measurement of the magnetic field is achieved by measuring the frequency and/or the voltage shift. Several magnetometers using the capacitive method have been presented in the literature; most of them were designed to improve the magnetometer performance by using different types of capacitance structures, and different ways of detection, including in-plane and out-of-plane detection, have been demonstrated. The next section summarizes some of these magnetometers and their performance.

3.2.1 In-Plane Capacitive Detection

Magnetometers based on in-plane capacitive detection are usually based on interdigitated fingers; moving and fixed fingers to detect the magnetic field. These types of magnetometers require a complex fabrication process [13].

During 2000, Emmerich proposed a new prototype shown in Figure 3.1. This proposed sensor has capacitive finger–type electrodes defined on a moving beam (moving electrodes) that is perpendicular to the excitation path and stationary finger–type electrodes that are bonded to the substrate (fixed part) [14]. Due to the presence of the Lorentz force, the moving beam position varies, leading to a capacitance change proportional to the applied magnetic field. First mode of resonance has been used to achieve maximum displacement. The sensor was fabricated using Bosch's standard surface micromachining process, which is used in many sensors such as accelerometers [15,16] and gyroscopes [17,18]. The fabrication process starts with a thermally grown oxide on a silicon substrate followed by a polycrystalline silicon layer, which is deposited on top of the oxide and then doped and patterned. Then, a sacrificial oxide layer is deposited and patterned, and a thick polysilicon (epi-poly) is grown and doped. Finally, as a conductor layer, aluminum is deposited on top of the stack followed by trench etching of the epi-poly layer prior to the release of micromechanical structures by etching the sacrificial oxide through holes using hydrofluoric acid (HF) vapor. The reported sensitivity value of the Emmerich magnetometer was 820 µV/µT, with a minimum resolution of 200 nT at a bandwidth of between 1 and 10 Hz. The resonant

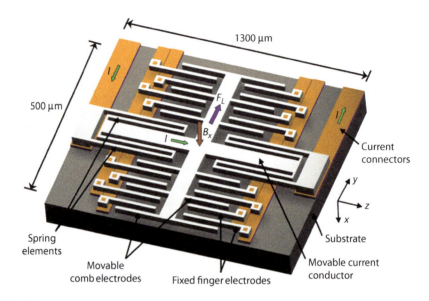

FIGURE 3.1
Schematic and operation principle of Emmerich resonator. (From Manjarrez, E. et al., *Sensors Journal*, 39(9), 7785–7813, 2009; Emmerich, H. and Schofthaler, M., *IEEE Transactions on Electron Devices*, 47, 972–977, 2000.)

frequency of the resonant structure was found to be 1.3 kHz and the quality factor was estimated to be 30 at 101 Pa [14].

In 2007, Bahreyni and Shafai designed a new resonant magnetometer that measures the frequency variation using in-plane electrostatic detection [19]. The sensor, shown in Figure 3.2, contains a shuttle-type structure connected to two bars using four microbeams. The shuttle connection is driven and kept in resonance by means of electrostatic actuation and sensing. The interaction between a dc drive current flowing in both bars and a magnetic field normal to the plane of the sensor, gives rise to a Lorentz force. This force produces axial stress on microbeams, which changes the resonance frequency of the comb. The shift in resonant frequency is controlled using capacitive detection with signal processing electronics. This device has a resolution of 217 nT and a sensitivity of 69.6 Hz/T for a current of 10 mA. The resonance frequency is around 27 kHz with a quality factor of 15,000 at 2 Pa. It should be mentioned here that the sensor's performance can be affected by the self-heating effect caused when a current is flowing through the microbeams [19].

In 2009, Thomson and Harsley [20] proposed a prototype that is quite similar to the Bahreyni et al. device. The main difference is related to the geometry of the structures and their shapes. A scanning electron microscope (SEM) photo of Thomson and Harsley's sensor is shown in Figure 3.3a. It consists basically of interdigitated fingers and a microbeam for the actuating current. An out-of-plane magnetic field produces the in-plane Lorentz force modulated at the device's natural frequency using an ac passing through the

Out-of-Plane Resonant Capacitive Detection Magnetometer

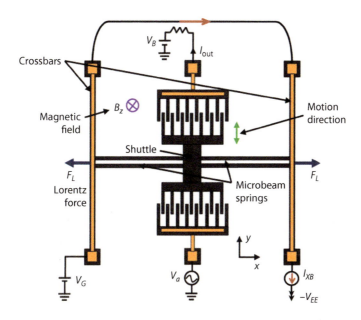

FIGURE 3.2
Schematic and operation principle of the Bahreyni resonator. (From Manjarrez, E. et al., *Sensors Journal*, 39(9), 7785–7813, 2009; Bahreyni, B. and Shafai, C., *IEEE Sensors Journal*, 7(9), 1326–1334, 2007.)

flexures. The device is parametrically driven by electrostatically modulating the mechanical spring stiffness at twice the natural frequency using a pump voltage applied to capacitive plates that are defined symmetrically around the device. This microsensor has a sensitivity of 31.4 mV_{rms}/T, a resonant frequency close to 8.46 kHz, and a quality factor of 48.8 at ambient pressure [20].

During the same year of 2009, Brugger and Paul developed a novel magnetic field microsensor combining both electrostatic and magnetic effects (see Figure 3.3b) [13]. A micromechanical resonating structure that is suspended by four straight flexural springs is electrostatically driven through interdigitated combs. The device is driven by applying a dc and a superimposed ac voltage to half of the interdigitated combs while the other half is used for the capacitive detection. In this sensor, the magnetic effect was monitored using a concentrator and a pair of planar coils. The concentrator is divided into three parts separated by two narrow gaps. The inner segment is attached to the resonator while both outer parts are fixed. The magnetic concentrator saturates at a magnetic field over 713 µT. The planar coil can add an auxiliary magnetic field parallel to the magnetic concentrator when it is actuated by an external current. Because of the magnetic field created by both the coil and the concentrator, a magnetic force counteracts the restoring force that is exerted on the resonator and affects its spring constant inducing a shift in the frequency of the resonance value. Finally, a relationship between the resonant frequency and the magnetic field can be obtained. For a current of 80 mA flowing

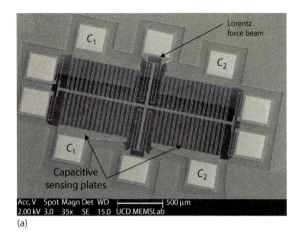

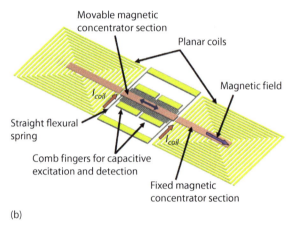

FIGURE 3.3
(a) SEM image of Thompson resonator. (From Thompson, M.J. and Horsley, D.A., *IEEE Sensors Conference*, 992–995, 2009.) (b) Operation principle of a magnetic field microsensor based on a magnetic concentrator developed by Brugger and Paul. (From Herrera-May, A.L. et al., Development of resonant magnetic field microsensors: Challenges and future applications, in *Microsensors*, I. Minin (Ed.), Chapter 3, InTech, Rijeka, Croatia, 2010; Brugger, S. and Paul, O., *Journal of Microelectromechanical Systems*, 18(6), 1432–1443, 2009.)

through the coil and with $V_{dc} = 20$ V and $V_{ac} = 404$ mV and under a pressure of 10^{-5} mbar, the microsensor presents a resolution of 1.3 µT with a sensitivity of 1.91 MHz/T, and a quality factor around 2400. It should also be mentioned here that the resolution of the sensor can reach 400 nT in vacuum [13].

3.2.2 Out-of-Plane Capacitive Detection

This section describes MEMS-based resonant magnetic field microsensors that use out-of-plane capacitive sensing. In 1998, Kadar developed a resonant magnetic field sensor with an out-of-plane capacitive detection scheme [21]. The microsensor combines bipolar processing with a micromachined structure that is obtained by glass to silicon anodic bonding. The structure is illustrated in Figure 3.4a. It comprises a resonant silicon plate with dimensions 2800 × 1400 µm with a rectangular coil deposited on top of it. When a sinusoidal current is passed through the coil, it interacts with the external magnetic field and a Lorentz force arises causing the motion of the moving plate. This movement induces a variation in capacitance between the moving structure and electrodes located on the packaging of the device. Then, an external circuit is used to convert the capacitance variation into induced voltages. This magnetometer has a sensitivity of 500 µV/µT and a resolution of 1 nT. The resonant frequency is close to 2.4 kHz with a quality factor of 700 at 5 Pa and the power consumption is a few milliwatts. The microsensor

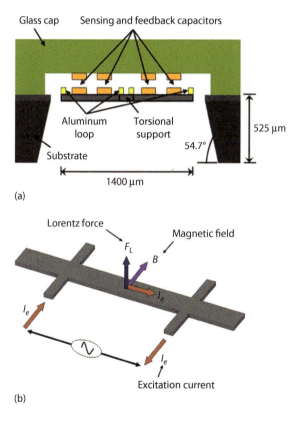

FIGURE 3.4
(a) Schematic of Kadar magnetometer. (From Manjarrez, E. et al., *Sensors Journal*, 39(9), 7785–7813, 2009; Kádár, Z., Bossche, A., Sarro, P.M. and Mollinger, J.R., *Sensors and Actuators A Physical*, 70(3), 225–232, 1998.) (b) Operation principle of a magnetic field microsensor developed by Tucker. (From Herrera-May, A.L. et al., Development of resonant magnetic field microsensors: Challenges and future applications, in *Microsensors*, I. Minin (Ed.), Chapter 3, InTech, Rijeka, Croatia, 2010; Tucker, J. et al., *2000 NanoTech*, Houston, TX, 1–5, 2002.)

requires a complex electronic circuitry for signal processing and vacuum packaging.

In 2002, Tucker designed a new structure for an out-of-plane capacitive detection (see Figure 3.4b) [22]. The structure consists of an xylophone microbar, supported by four arms, to detect external magnetic fields. When a sinusoidal current flows through the microbar, which is submitted to an external magnetic field, a Lorentz force is generated that causes the microbar to vibrate. This vibration is measured as a capacitance variation. The magnetometer area is close to 0.5 mm² and was fabricated using a complementary metal-oxide semiconductor (CMOS) 0.5 μm standard commercial process. A post-process release step was added to release the microbar. The sensor has a resonant frequency of around 100 kHz with a quality factor of 1000. The power consumption is around 7.5 mW and the noise is measured to be $0.5\,\text{nT}/\sqrt{\text{Hz}}$. In order to increase the vibration amplitude of the microbar, the device has to be packaged in vacuum.

In 2008, Kyynäräinen et al. fabricated a magnetometer, shown in Figure 3.5, which is based on an out-of-plane capacitive detection scheme [23] between a square coil on top of a moving structure (electrode) and a fixed electrode. When current flows through the coil in the presence of an external magnetic field in the y-axis direction, an out-of-plane Lorentz force appears and the movement of the structure gives a capacitance variation between the electrodes that is proportional to the component of the magnetic field in the y-axis. The sensor is fabricated with direct bonding of a double-sided polished silicon wafer and a silicon on insulator (SOI) wafer. The main steps in the manufacturing process of this magnetometer are patterning of the anti-stiction studs and metalizing of the coil metal compensation layers, cavity etching and metalizing of sensing electrodes for torsional sensors, and direct wafer bonding followed by the removal of the SOI wafer handle layer. The excitation coil is patterned next, followed by the deposition of the isolation and the bridge metal layers and then the final release etch step. The magnetometers operate in vacuum to increase sensitivity readings. The measurement of the field component along the chip surface gives a flux density resolution of about $10\,\text{nT}/\sqrt{\text{Hz}}$ at a coil current of 100 μA.

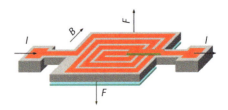

FIGURE 3.5
Schematic of Kyynäräinen magnetometer design. (From Kyynäräinen, J., *Sensors and Actuators A*, 142(2), 561–568, 2008.)

Magnetometers measuring the field component perpendicular to the chip surface are less sensitive with a flux density resolution of about 70 nT/$\sqrt{\text{Hz}}$.

Another alternative structure was proposed by Ren in 2009 [24]. Similar to the previous techniques but with different geometry, the prototype has a low resistivity silicon structure suspended by two torsional beams over a glass substrate. On top of the silicon, a thin layer coil with a thickness of 1 µm and a width of 0.5 µm is deposited for excitation. When feeding the coil with an ac in the presence of an external magnetic field, the suspended structure starts oscillating, producing a capacitance change between the silicon resonator and the capacitance plate (see Figure 3.6). A capacitance detection circuit measures the capacitance change, which actually depends on both the magnitude and the direction of the external magnetic field. The magnetometer has been microfabricated using conventional MEMS micromachining and silicon-to-glass anodic bonding. For a 150 mV driving voltage, the microsensor shows an improved performance with sensitivity equal to 481 mV/T, for a magnetic field varying from 3 to 30 µT. The resolution is estimated to be 30 nT and the resonant frequency is close to 1380 Hz, with a quality factor of 2500 at a pressure of 10 Pa. This type of sensor has a nonlinear response from 0 to 3 µT.

Generally, a resonant magnetic field microsensor with capacitive detection needs vacuum packaging in order to increase the quality factor, thereby enhancing the displacement of the resonant microstructure. Therefore, both the sensitivity and the resolution of the device will increase [11]. The quality factor, Q, is important for the resonant microstructure performance and is closely related to the damping effect. This factor is expressed as

$$Q = 2\pi f_0 \frac{m}{\alpha} \quad (3.1)$$

where:
- f_0 is the structure resonant frequency
- m is the mass
- α is the damping coefficient

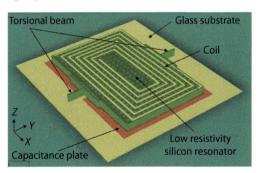

FIGURE 3.6
Operation principle of a magnetic field microsensor developed by Ren. (From Ren, D., *Sensors Journal*, 9(9) 6951–6966, 2009.)

In reality, energy is dissipated during the structure's motion and the magnitude of the oscillation decreases until the oscillation stops. This energy dissipation is known as *damping*. If an undamped structure is allowed to vibrate freely, the amplitude of vibration will be infinite. The damping level of a system is determined by the quality factor, which measures the amount of energy loss per cycle during the operation of the resonant structure. Then, due to the damping effect, the total Q factor can be written as the ratio of the total energy that is stored in the structure (E_M) over the energy loss per cycle (E_C) [25]:

$$Q = 2\pi \frac{E_M}{E_C} \tag{3.2}$$

The energy loss can be due to different mechanisms. Some of these mechanisms are intrinsic to the material (thermoelastic damping), while others are influenced by the geometry of the structure, or by interactions with the environment in which the solid body vibrates (air damping, squeeze film damping) [26]. As a result, quality factors depend on the fluid type and pressure, the vibration mode of the structure, the size and shape of the resonant structure, and the distance (gap) with respect to adjacent surfaces. The MEMS designer needs to carefully consider this parameter in order to increase the magnetometer's performance. For example, many magnetometer designs use a perforated microstructure to decrease the damping effect [27]. Since for most MEMS operating in air, the main energy loss is air damping, operating at lower pressures or in vacuum increases the quality factor; however, in this case, current-driven devices such as the Lorentz force–based magnetometers may be severely impacted by joule self-heating effects [26]. Finally, another important drawback of this type of microsensor is that it suffers from parasitic capacitances, which can be reduced through the monolithic integration of the micromachined resonators with electronic circuitry. In the next section, the design for the developed out-of-plane capacitive detection will be described and analyzed. The prototype was designed at UCL in Belgium.

3.3 Magnetometer with Capacitive Detection

A schematic of the xylophone-based magnetometer with capacitive detection is shown in Figure 3.7. In this sensor, the out-of-plane displacement of a resonating beam structure affects the capacitance between interdigitated fingers defined on the two sides of the beam and the substrate [28,29]. The contribution of our design is the capacitance detection scheme, which is different from previous designs of sensors using out-of-plane detection [22].

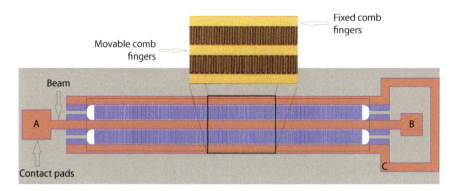

FIGURE 3.7
Top view schematic of the xylophone-based magnetic field sensor with capacitive detection.

The design of the magnetometer consists of a movable clamped-clamped beam with fixed comb electrodes defined along its entire length. These electrodes form interdigitated pairs with combs attached to the substrate on both sides of the resonant beam. Then, the sensor is placed in an external static magnetic field. According to the Lorentz force law, when a current passes through a suspended beam (between pads A and B in Figure 3.7) and is coupled to the external magnetic field, a Lorentz force arises on the beam that is oriented according to the right-hand rule (perpendicular to both the magnetic field and the current). The clamped-clamped beam will thus move out of the plane (z-direction) when the magnetic field is oriented in the plane. As a result, the capacitance between the interdigitated electrodes changes and can be measured between pads A and C. In this design, the beam is made long enough to accommodate a relatively high number of comb fingers on both sides. This will eventually produce a more pronounced variation in the capacitance formed by the interdigitated combs, making it easier to measure. The choice of beam length is restricted by limitations on the maximum area that can be etched from the substrate, which are imposed by the fabrication technology. On the other hand, the beam's width should be minimized to reduce energy loss due to air damping. The width of the capacitive fingers on each side of the beam has been chosen small enough to maximize the number of fingers along the beam length. In addition, the fingers are made long and are located very close to each other to increase the capacitance variation. The fingers' thickness depends on the beam and is imposed by the fabrication process. So, the magnetometers come in two versions: "thin" or "thick," with 10 and 20 μm thick Si structural layers, respectively. The optimized geometrical parameters that were used for the fabrication of our magnetic field sensor are summarized in Table 3.1. The sensor has been fabricated using the silicon-on-insulator multiuser MEMS process (SOIMUMPs) foundry process (MEMSCAP Inc., Durham, North Carolina) with a silicon thickness equal to 10 μm.

TABLE 3.1
Used Optimized Dimensions for the Magnetic Field Sensor Design

Part	Parameters Designation	Value (µm)
Beam	Length (L)	1969
	Width (W)	62.5
	Height (h)	10
Fingers	Length (l)	100
	Width (a)	3
	Thickness (t)	10 and 25
	Fingers number (n_t)	312
	Gap between fingers (d)	3

3.4 Mechanical and Electrical Behavior

3.4.1 Magnetometer Mechanical Model Analysis

The key mechanical parameter of the magnetometer is the resonant frequency of the clamped-clamped beam. In practice, the sensor is placed in a 150 mT static magnetic field (generated by Nd–Fe–B magnets). The beam is driven by an ac with a frequency equal to its first resonant frequency to ensure the highest amplitude of vibration. The beam eigenfrequencies are mainly determined by the material properties and the dimensions of the structural layer. The resonance frequency, f_n, of a given vibration mode n of the clamped-clamped resonant beam, shown in Figure 3.6, is obtained by [30]

$$f_n = \frac{R_n^2}{2L^2} \sqrt{\frac{EI}{\rho A}} \tag{3.3}$$

where:
- ρ is the density of the structural material
- A is the beam cross-sectional area
- E is Young's modulus
- R_n is a constant that depends on the boundary conditions and the natural mode of the beam (values are given in Table 3.2)
- I is the area moment of inertia given, for a rectangular section beam, by $W.h^3/12$

Based on the values given in Table 3.1, the frequency, f_0, of the first resonant mode is calculated from Equation 3.3 and is equal to 17.5 kHz.

The resonant frequency and the displacement of the clamped-clamped beam for the first mode of vibration are measured using laser Doppler velocimetry (LDV; Polytec MSA-500), as shown in Figure 3.8. In this setup, the

TABLE 3.2
R_n Constant Values for the First Five Eigenmodes

Parameter	Value
R_1	4.738
R_2	7.853
R_3	10.995
R_4	14.131
R_5	17.278

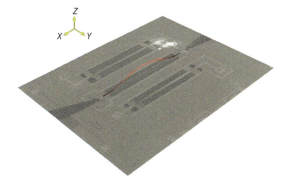

FIGURE 3.8
Out-of-plane displacement of the resonant clamped-clamped beam illustrated using Polytec MSA-500.

beam is actuated first by a periodic chirp signal of fixed amplitude that is applied between pads A and B (see Figure 3.7) in order to determine the resonant frequencies. Based on this experiment and referring to the obtained spectrum shape, the resonant frequency was measured and was equal to 18.8 kHz. The difference between experimental and theoretically obtained values is because the assumed model is a simple silicon beam without fingers and without the Au on its top layer.

The response of the sensor is defined by the displacement of the vibrating beam. In order to measure this using LDV, a pure sinusoidal current with a frequency equal to the first resonant frequency of the resonating mechanical structure is used to drive the beam. Figure 3.9a shows the displacement of the beam center at resonance as a function of the root mean square (rms) value of the actuating current. It is worth noting that the variation in the maximum displacement is almost linear between 0 and 48 mA rms. Figure 3.9b shows the normalized modal shape of the beam when actuated by a current at its first resonant mode. The amplitude of vibration was measured to be close to 3.8 µm in the beam's center when an rms current of 48 mA is flowing through the beam.

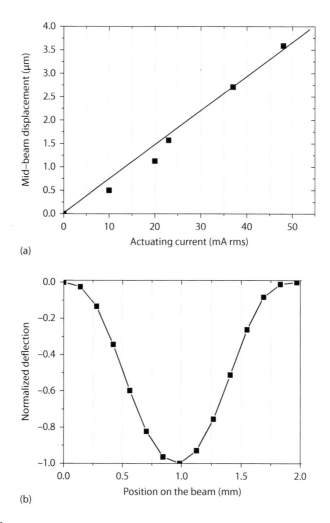

FIGURE 3.9
(a) Measurement of the central beam displacement as a function of the supply current. (b) Normalized beam deflection shape when actuated at resonance frequency.

The measured resonant frequency is plotted in Figure 3.10 as a function of the supply current. It can be immediately seen that the resonant frequency shifts slightly from its initial value when the actuating current increases. We note here that contrary to this observation when testing our thicker structures (25 μm thickness), we observe small shifts of the resonant frequency with current toward smaller values. This is expected because during operation, the temperature of the device rises due to the joule heating effects, causing a compressive internal stress to arise and lowering the resonant frequency [29]. As the excitation current increases, the temperature rise is intensified resulting in a higher compressive stress and larger shifts. In this

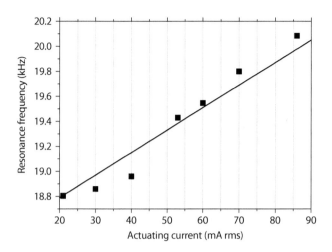

FIGURE 3.10
Measured resonance frequency variation as a function of the supply current.

case, however, the shift to higher frequencies suggests that another physical mechanism is present, especially at higher currents, and dominates the impact of the compressive thermal stress. We currently suspect that the vibration of the thinner sensors becomes nonlinear as the current exceeds a certain value and this causes the observed shift in the first resonant frequency with the current. Based on Figure 3.10a, a linear fit was plotted and an analytical expression was developed between the measured resonance frequency and the used rms current, J, which can be given by

$$f_1 = f_0 + 0.0193\ J \tag{3.4}$$

The fact that the resonant frequency of the beam shifts due to thermal and nonlinear effects can cause a serious dysfunction during the operation of our magnetometer that may affect its response and reduce its sensitivity. Consequently, there is a need to accurately tune the microsensor and this will be investigated hereafter.

3.4.2 Electric Model Analysis

The parameters of the sensor electric model are deduced from the measurements of the input and output impedances. Both of these quantities are measured using the Agilent 4284A LCR meter in a frequency range of between 20 Hz and 1 MHz. The input impedance of the sensor is equivalent to a resistance formed by the metal layer that is deposited on top of the beam. Experimentally, this impedance is measured between pads A and B (see Figure 3.7). In Figure 3.11, the input impedance and the corresponding

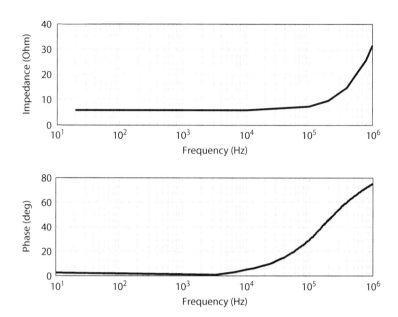

FIGURE 3.11
Experimental measurement of the magnetometer input impedance.

phase as a function of frequency are plotted for a 10 μm-thick structure. From these curves, the resistance of the Au layer was deduced and is equal to 5.62 Ω for a frequency less than 100 kHz, after that an inductive effect appears. The output impedance of our structure is also important because it is defined by the total capacitance of the finger electrodes. Experimentally, this impedance is measured between pads A and C (see Figure 3.7). From the schematic in Figure 3.12 (modulus and phase of output impedance), the impedance reduces with frequency, which clearly explains the effect added by the capacitance. A direct measurement with an LCR meter also shows the presence of a serial resistance with the capacitance in this output impedance. The values obtained from these measurements at different frequencies are summarized in Tables 3.2 and 3.3.

The magnetometer electric model can be represented as shown in Figure 3.13, where $R_{\_act}$ represents the input impedance. The impedance $(C_{\_eq}, R_{\_eq})$ models the equivalent circuit of the RC dipole, which was derived from the output impedance. The beam is actuated between A and B pads and the capacitive measurement is measured between A and C pads or B and C pads. For high frequencies, the input impedance can change in the electric model.

In the next section, the sensitivity of the magnetometer will be evaluated. To do this, we need to group the different domains involved: magnetic, mechanical, and electrical domains.

Out-of-Plane Resonant Capacitive Detection Magnetometer

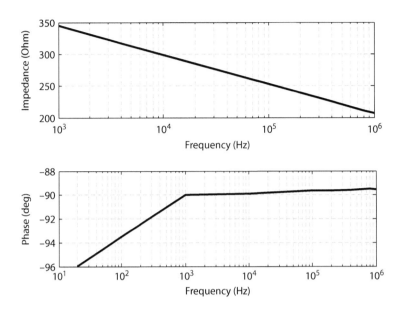

FIGURE 3.12
Experimental measurement of the magnetometer output impedance.

TABLE 3.3

Capacitance and Resistance Values Obtained from Output Impedance Measurements at Different Frequencies for 10 mm-Thick Structure

Frequency (kHz)	Capacitance (pF)	Resistance (MΩ)
1	6.76	>200
100	6.72	>200
1000	6.61	>200

3.4.3 Magnetometer Sensitivity Evaluation

To determine the electrical properties of the sensor, we calculate the change in the capacitance for different beam displacements. Since the beam is moving in the out-of-plane direction (z-axis), the magnetometer is more sensitive along the z-axis direction. The Lorentz force applied on the beam, F_L, can be given by

$$F_L = B_y L J \tag{3.5}$$

where:
    B_y is the magnetic flux density in the y-direction
    J is the current flowing in the beam in the x-direction

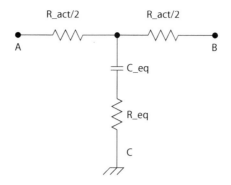

FIGURE 3.13
Electric model of the magnetometer sensor.

The deflection profile of a fixed-fixed beam submitted to a uniform load given by Equation 3.5, expressed by J, B_y, throughout its length is given by

$$z(x) = B_y J \frac{x^2}{24EI}(L-x)^2 \qquad (3.6)$$

where x is the direction across the beam length (see Figure 3.7). To discretize this equation, we can write that $x = n \times (d + a)$, where n denotes the rank of the finger, thus $z(n)$ becomes the out-of-plane displacement of the nth finger. The theoretical expression of the capacitance in this configuration is given by [31]

$$C(n) = \begin{cases} n_t \varepsilon_0 \dfrac{l(t - |z(n)|)}{d}, & \text{if } |z(n)| < t \\ 0, & \text{if } |z(n)| > t \end{cases} \qquad (3.7)$$

where:
  ε_0 is the relative permittivity of the air
  n_t is the number of comb capacitance fingers

The others parameters are given in Table 3.1. In order to accurately evaluate this capacitance, a finite element method (FEM) simulation was carried out using Comsol® Multiphysics. Firstly, a single interdigitated finger pair was studied and simulated using an electrostatics module. This was accomplished by applying a dc voltage between the finger pair while using silicon as the material and air in the gap. Then, the distance was varied between the movable and the fixed fingers along the z-direction, and we tried to determine numerically the capacitance between the finger pair. Thereafter, two methods were developed and compared to estimate the total capacitance. The first method is calculating the capacitance between two fingers pair as described, then multiplying the result by the total number of finger pairs throughout the clamped-clamped beam n_t; the result is plotted in Figure 3.14a. The second method is more realistic, and takes into consideration the edges effect during the beam motion, so that the elevation is not the same throughout its length, as show in Figure 3.9b. The partial capacitance produced by each comb fingers throughout the clamped-clamped beam length is presented in Figure 3.14b. The simulated resulting total capacitance variation as a function of the gap between fingers, for both methods, is plotted in Figure 3.14a. It clearly shows that the capacitance value is maximum at the equilibrium position, then decreases as the beam moves away from its rest position.

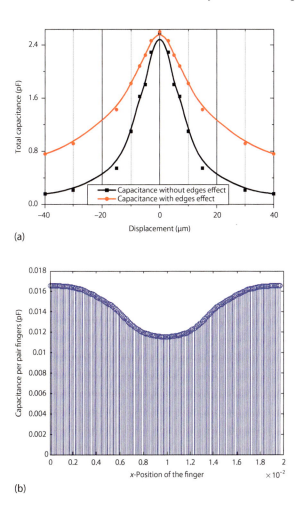

FIGURE 3.14
(a) Total capacitance change as a function of the displacement between the comb fingers. (b) Partial capacitance produced by each comb finger throughout the clamped-clamped beam length.

A key parameter for the function of the microsensor is sensitivity, which can be defined by Equation 3.8 for a constant static magnetic field:

$$S = \left.\frac{\Delta C}{\Delta J}\right|_{B=cte} = \left.\frac{\Delta C}{\Delta z}\frac{\Delta z}{\Delta J}\right|_{B=cte} \tag{3.8}$$

The sensitivity equation can be divided into two parts as shown in Equation 3.8, and can be determined graphically. In fact, the first part can be deduced from Figure 3.14a, whereas the second part can be extracted from Figure 3.9a, which presents the displacement as a function of the actuating current. In Figure 3.15a and b, respectively, we plot the change in total

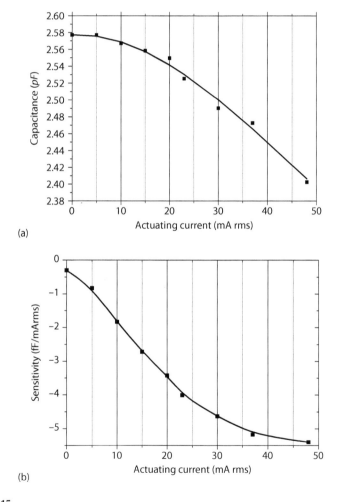

FIGURE 3.15
Simulated (a) capacitance and (b) sensitivity of the magnetometer as a function of the actuation current.

capacitance and the sensitivity for different actuating currents. The capacitance is maximum for low actuating currents (low displacement) while the sensitivity in that case is lowest. For a current of 48 mA rms and when a static magnetic field of 150 mT is applied, the sensitivity is around −7 fF/mA rms, as shown in Figure 3.15b. The maximum sensitivity is found to be around −46.6 fF/T mA rms.

The sensor sensitivity depends mainly on the chosen geometrical parameters during the SOIMUMPs fabrication process and on the quality factor [18]. The sensor performance can be improved when using a vacuum package. Indeed, in vacuum both the quality factor and the amplitude of vibration increase by reducing the main energy losses. It may be noted that the sensitivity is also directly proportional to the current passing through the beam; however, increasing this current causes a shift of the resonant frequency as shown in Figure 3.10, which presents a drawback for the present resonant sensor design. Again, the implementation of a tuning system to solve this problem becomes a necessity. In the next section, a proposed tuning technique will be discussed using the electrostatic actuation force between fingers.

3.5 Frequency Tuning by Applying dc Voltage

In practice, the measured resonant frequencies always shift from the desired design values because of several mechanisms, such as joule heating, external temperature influence, and external vibration. Two solutions could be conceived to compensate for the frequency deviation: widening the operating frequency bandwidth and/or tuning the resonant frequency. The first method can be implemented by adjusting the quality factor of the vibrating structure as mentioned in [32]. The quality factor is tuned using an annealing technique that is based on filament-like heating of our microstructure. However, the resonant frequency tuning method is a more efficient technique for applications with single time-varying dominant frequency [32]. Frequency tuning could be achieved by one of three basic methods: the first consists of using an additional electrostatic stiffness component, wherein the stiffness can be adjusted electrostatically through an interdigitated fingers capacitive structure. In fact, Lee et al. [33] demonstrated that by inducing a control voltage between microresonator comb fingers when structures are actuated laterally, the resonant frequency decreases. They show that this decrease is due to the stiffness from the electrostatic force component that actually reduces the total effective stiffness of the structure [33]. So, design dimensions were chosen to generate higher electrostatic stiffness to get a wide tuning frequency variation. The second method exploits the temperature dependence of Young's modulus during the elastic deformation of some materials. It is well known that Young's modulus of silicon, for instance, depends on temperature. Therefore, any temperature change can cause a shift in the stiffness of any silicon-based structure, which in turn affects its initial resonant frequency [25]. The last method is based on tuning by inducing an internal stress in the structure (compressive or tensile) in order to cause shifting of the resonance frequency. Todd and Liwei [34] used the joule heating effect by applying a current to the comb structure through a straight beam designed with specific dimensions and materials to induce thermal stress. The resonant frequency was then changed by around 6.5% for a tuning current of 5 mA [34]. The method presented in this chapter exploits the former principle. During the out-of-plane movement of the clamped-clamped beam, its resonant frequency was tuned via the electrostatic method by applying a dc voltage between its capacitive fingers. For this magnetometer design, the resonant frequency increased with increasing dc voltage. This is different from other models described in the literature since the electrostatic force is produced in the z-direction and not laterally [35,36].

3.5.1 Physical Modeling

In Figure 3.16a, a 3-D schematic of a sensor with comb fingers is depicted. The two components of the electrostatic force at equilibrium, across the x- and y-directions, are also shown. However, during the out-of-plane motion, the stationary electrodes are positioned exactly in the middle of the movable finger, so the in-plane electrostatic force components (x and y) from both sides are opposed and cancel each other out, as shown in Figure 3.16a. Thus, only when the beam starts oscillating out of plane will the z-axis component of the electrostatic force rise. This force, F_{ez}, can be evaluated through the electric energy using [12]

$$F_{ez} = \left.\frac{\partial w_e(z)}{\partial z}\right|_{V=cte} \tag{3.9}$$

where w_e is the electrostatic energy stored between the finger electrodes when a dc voltage V is applied, and can be expressed by

$$w_e = \tfrac{1}{2} C(z) V_{dc}^2 \tag{3.10}$$

The total electric energy is presented in Figure 3.16b for different bias voltages over the finger's displacement. As expected, Figure 3.16b shows that the electrical energy is at its maximum in the vicinity of the equilibrium position, then decreases as the beam moves away.

The moving part of the sensor (blue in Figure 3.16a) is modeled as a simple clamped-clamped beam with both ends fixed to the substrate. The effects of build-in and thermal stresses, which may rise due to joule heating of the Au electrodes, are ignored. It is further assumed that the geometry of this beam is simple with no comb fingers. However, the presence of the comb fingers

Out-of-Plane Resonant Capacitive Detection Magnetometer

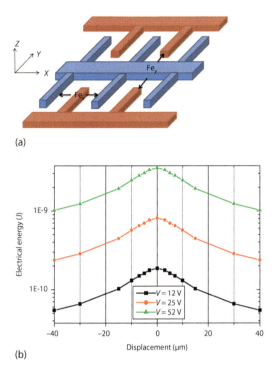

FIGURE 3.16
(a) 3-D schematic of the sensor with electrostatic force distribution in equilibrium ($z = 0$) between two fixed combs when a dc voltage is applied. (b) Total electric energy over the finger's displacement for different bias voltages.

alters the geometry and the flexural rigidity of the structure. In order to take this into account, we introduce an equivalent width, b_{eq}, as the width of a beam with a constant cross section having the same length and thickness as our sensor, which—assuming pure bending conditions—will deform with the same curvature under the action of an external bending moment M. This is mathematically formulated in Equation 3.11, when assuming a homogeneous beam with constant Young modulus E, given by

$$\int_0^L \frac{M}{EI_{eq}} dx = \int_0^L \frac{M}{EI(x)} dx \Rightarrow \frac{L}{I_{eq}} = \int_0^L \frac{dx}{I(x)} \quad (3.11)$$

The integral in Equation 3.11 is transformed to a sum of terms and the equivalent area moment of the beam, I_{eq}, is obtained. From this, an equivalent width can be calculated. We further take into account the bilayer structure of the sensor by calculating the flexural rigidity of the bilayer beam formed by Si–Au, which is given by

$$(EI)_{bilayer} = E(\text{Si})I_{eq}(\text{Si}) + E(\text{Au})I_{eq}(\text{Au}) \quad (3.12)$$

The Euler–Lagrange dynamic beam equation, describing the vibrations of a beam subject to an external load q, is given next. In Equation 3.12, x-coordinate is chosen across the direction of the beam length. The out-of-plane deflection at the z-direction in position x at time t is given by $z(x,t)$. Constant α is the damping ratio and p is the mass of the beam. Equation 3.13 was formulated making the usual assumption that bending effects are much more important than tension effects in a clamped-clamped beam.

$$\frac{\partial^2}{\partial x^2}\left(EI\frac{\partial^2 z}{\partial x^2}\right) - \alpha\frac{\partial z}{\partial x} = \mu\frac{\partial^2 z}{\partial x^2} + q(x,t) \tag{3.13}$$

In the case of our sensor, external load variable q can be divided into two discrete components. The first is the actuating external force, $F_L = BJL\cos(\omega t)$, due to the Lorentz interaction between the ac and the external magnetic field $\vec{B}$, which is applied to the midpoint of the suspended structure. The other contribution to q is the electrostatic interaction between the comb fingers at the two sides of the vibrating beam with the stationary comb fingers on the substrate. This electrostatic force is always attractive and is assumed that it is homogeneously distributed along the length of the beam.* We next attempt to calculate the electrostatic attraction that is exerted on each comb finger as a function of the tuning dc voltage (V_{dc}). The governing equation of motion of a single finger moving in the electrostatic field generated by a pair of stationary fingers on the substrate is given by

$$\mu_f\frac{\partial^2 z}{\partial t^2} = -\frac{\varepsilon_0}{2}|\nabla\Phi|^2 = -\frac{\varepsilon_0}{2}\left[\left(\frac{\partial\Phi}{\partial z}\right)^2 + \left(\frac{\partial\Phi}{\partial x}\right)^2\right] \tag{3.14}$$

In Equation 3.14, elastic and damping effects are ignored. It is thus assumed that the fingers are 2-D bodies that do not deform and vibrate without losses. The right-hand side of Equation 3.14 corresponds to the electrostatic force due to the presence of a potential Φ, ε_0 is the dielectric permittivity of vacuum, and μ_f is the surface mass density of one comb finger. In Figure 3.17, a cross-sectional view of a pair of stationary comb fingers at constant potential V_0 and a floating finger vibrating between them is schematically depicted. This is a simplified model of a comb finger moving as a dielectric slab between the two stationary fingers that form a capacitor. In reality, we need to take into account the contribution of neighbor fingers in the potential Φ. However, we can ignore this effect for this first approximation of the electrostatic force on one comb finger. In Figure 3.17, the main dimensions that are relevant to this problem are also drawn. In what follows, we scale the terms in Equation 3.14 using appropriate non-dimensional parameters for x, z, t, and Φ in order to

---

* This is a simplification that is not expected to affect our results because the comb fingers are defined across the central section of the beam spanning a length of 1860 μm, which is slightly shorter than the beam's length.

Out-of-Plane Resonant Capacitive Detection Magnetometer

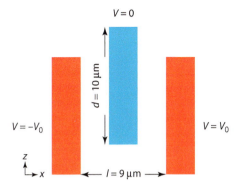

FIGURE 3.17
In-plane schematic representation of comb fingers.

draw important conclusions about the physical characteristics of the system described by Equation 3.14. By choosing d as the length scale for both x and z since $x \approx z$, V_0 as the scale for potential Φ, and T as the scale for time t, we get the non-dimensional primed parameters $x' = x/d$, $z' = z/d$, $\Phi' = \Phi_0/V_0$, and $t' = t/T$.

By transforming the partial derivatives using the chain and substituting into Equation 4.14, we get the non-dimensional form:

$$\mu_f \frac{d}{T^2} \frac{\partial^2 z'}{\partial t'^2} = -\frac{\varepsilon_0}{2} |\nabla \Phi|^2 = -\frac{\varepsilon_0 V_0^2}{2d^2} \left[\left(\frac{\partial \Phi'}{\partial z'} \right)^2 + \left(\frac{\partial \Phi'}{\partial x'} \right)^2 \right] \quad (3.15)$$

It immediately follows from Equation 3.15 that the appropriate scaling factor T for time t is given by Equation 3.16. Here, the terms on the right-hand side can be reordered in the form m/k and the denominator $(\varepsilon_0 V_0^2 / 2d^3)$, which is a quadratic function of the external potential V_0, is the key factor that controls the "tuning" effect [37].

$$T = \sqrt{\frac{2\mu_f d^3}{\varepsilon_0 V_0^2}} = \sqrt{\frac{\mu_f}{\frac{\varepsilon_0 V_0^2}{2d^3}}} \quad (3.16)$$

3.5.2 Numerical Resolution

An analytic solution of the total electrostatic force exerted on a dielectric slab of length l, with dielectric permittivity ε, which is inserted in a parallel plate capacitor whose plates are kept at a distance d, exists and is well known in the literature as $F_{elec} 1 = (\varepsilon - \varepsilon_0) l \Delta \Phi^2 / 2d$, where $\Delta \Phi$ in this case is the potential difference between the parallel plates of the capacitor. However, the assumptions behind its derivation are not always clearly discussed and, as a result, it is usually used in the wrong context. For an in-depth discussion of the physical interactions involved, interested readers should refer to [38]. Under more general conditions, Equation 3.15 is not open to analytic solutions. We have thus resorted to finite element analysis in order to calculate the electrostatic force exerted on a comb finger–type electrode from a parallel plate capacitor. The z-component of the electrostatic force that is exerted on the device as a function of the out-of-plane displacement can be determined by differentiating the total electric energy of the system. The resulting electrostatic force values are plotted as a function of the displacement in Figure 3.18 (dashed line with circle dots) for the case $V_{dc} = 12$ V. This latter does not represent a linear function except in a narrow region around the equilibrium position. It assumes its maximum/minimum value in a distance approximately equal to 1.2 µm away from the rest point, after which it decreases exponentially to zero.

The electrostatic restoring force can be approximated by a function that should be almost linear close to zero and tends to die out at large distances away from the equilibrium position. Based on these two prerequisites, the function describing the total force exerted on the moving part, as a function of the displacement z, can be expressed as

$$F_{elec} = \beta z e^{-|z/\delta|} \tag{3.17}$$

We now return to Equation 3.13 and reformulate it in order to take into account the extra restoring force due to the electrostatic interaction. We ignore damping effects but we also take into account the equivalent area moment of inertia, I_{eq}, and the flexural rigidity of the bilayer structure, calculated from Equations 3.11 and 3.12, and we end up with

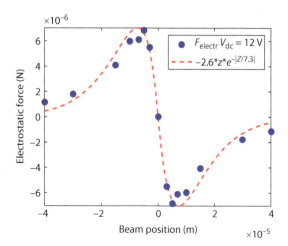

FIGURE 3.18
Representation of the electrostatic force exerted on a magnetometer (dashed line with solid blue circle), and the fitting function (red dashed line) obtained by applying a least squares fitting method, as a function of the out-of-plane displacement for $V_{dc} = 12$ V.

$$\mu\frac{\partial^2 z}{\partial t^2} + EI\frac{\partial^4 z}{\partial x^4} + \beta z e^{-|z/\delta|} = F_L \tag{3.18}$$

Next, the parameters used in Equation 3.18 are scaled. For this, the x-coordinate across the length L of the beam is scaled using L as the scaling parameter; w the coordinate in the direction of oscillation is scaled by d, which should correspond to a measure of the out-of-plane deflection; and T is a time parameter that is chosen equal to $L^2\sqrt{\mu/EI}$. By transforming the partial derivatives using the chain rule, we get the non-dimensional form:

$$\frac{\partial^2 z'}{\partial t'^2} + \frac{\partial^4 z'}{\partial x'^4} + \frac{\beta L^4}{EI}z'e^{-|dz'/\delta|} = \frac{L^4}{EId}F_L \tag{3.19}$$

The total force due to the Lorentz interaction is given by $F_L = BJL\cos(t')$. This force is applied at the midpoint of the beam when the excitation current with an amplitude J is used in the presence of an external magnetic field $\vec{B}$. The equivalent spring constant of a suspended clamped-clamped beam is given by $k_{eq} = 192EI/L^3$ when a point load is applied at the midpoint of the beam, choosing d as the ratio of static load over spring constant ($BJL/EI/L^3$) looks the most straightforward choice. This corresponds to the expected maximum deflection of the beam under the action of the Lorentz force. It is a function of the excitation current and of the external magnetic field B. We should further notice that the coefficient of the electrostatic term $\beta L/EI/L^3$ is the ratio of the stiffness due to the electrostatic attraction over the stiffness of the beam and represents, as a result, an indication of the relative importance of the two terms. From the foregoing explanations, Equation 3.19 becomes

$$\partial_{tt}z + \partial_{xxxx}z + Kze^{-|dz/\delta|} = \cos(t) \tag{3.20}$$

In Equation 3.20, we have dropped primed variable names for simplicity. We have also adopted a simpler symbol for partial derivatives over t and x and have introduced $K = \beta L^4/EI$ as the coefficient of the electrostatic term. If we further focus on the free vibrations of the structure and replace the stiffness term with the equivalent spring constant, we obtain

$$\partial_{tt}z + z + Kze^{-|dz/\delta|} = 0 \tag{3.21}$$

In the absence of an external electrostatic field, $K = 0$ and Equation 3.21 corresponds to the free vibrations of a 1-D harmonic oscillator with $\omega^2 = 1$. As V_{dc} increases, K also increases to 1 and beyond, making the electrostatic term increasingly important and even dominant. The natural frequency of the system is now given for small vibrations by $\omega^2 = 1 + K$ and the phase diagram of the oscillatory motion can be derived for different values of K.

Next, we investigate the modeling of the damping effects included in Equation 3.19. In this case, scaling of the time parameter t is done using $T = \mu/\alpha$ as the scale factor, where α is a damping-related parameter. Then, we obtain

$$\frac{\partial^2 z'}{\partial t'^2} - \frac{\partial z'}{\partial t'} + \frac{\mu EI}{\alpha L^4}\frac{\partial^4 z'}{\partial x'^4} + \frac{\beta \mu}{\alpha^2} z' e^{-|dz'/\delta|} = \cos(t') \quad (3.22)$$

For the derivation of Equation 3.22, we have made the choice of the scaling parameter d following the same logic as for Equation 3.19. In this case, we set $d = F_L \mu/\alpha^2$. Next, we simplify the format of Equation 3.22 by following the same steps as before (see derivation of Equation 3.21). Here, however, we set $K_1 = \mu EI/\alpha^2 L^4$ and $K_2 = \beta\mu/\alpha^2$ and we observe that K_2/K_1 is now equal to the coefficient K of the electrostatic term in Equation 3.21. Then, Equation 3.22 becomes

$$\partial_{tt} z - \partial_t z + K_1 \partial_{xxxx} z + K_2 z e^{-|dz/\delta|} = \cos(t') \quad (3.23)$$

An alternative way to scale Equation 3.22 is by introducing t' from Equation 3.19 instead of dividing by $T = \mu/\alpha$ as we did. In that case, we end up with the following form:

$$\partial_{tt} z - \gamma \partial_t z + \partial_{xxxx} z + Kz e^{-|dz/\delta|} = \cos(t') \quad (3.24)$$

In Equation 3.24, the constant K is defined in the same way as in Equation 3.20 and measures the relative strength of the electrostatic force: it is thus a measure of the tuning effect. Constant γ, which appears as the coefficient of the first derivative with respect to time, is the reciprocal of the quality factor Q of the system and is defined from

$$\gamma = \frac{\alpha L^2}{\sqrt{\mu EI}} \quad (3.25)$$

In order to get a non-dimensional force term in the right-hand side of Equation 3.24, we need to set d, which as before is used for scaling lengths in the out-of-plane direction, equal to $F_L L^4/EI$ or $d = \mu F_L \gamma^2/\alpha$, which is derived from solving Equation 3.25 for $L^4/EI = \mu\gamma^2/\alpha^2$.

Given that for these devices Q was obtained experimentally ($Q \approx 100$), we get $\gamma \approx 0.01$ and as a result of Equation 3.24, $\alpha \approx 10^{-5}$. The partial differential of Equation 3.24 was numerically solved in MATLAB©. The solution is calculated for three units T of time and the resonance frequency is extracted from the period of the oscillation of the beam.

In Figure 3.19, the numeric solutions obtained from MATLAB for a driving current with amplitude $J = 10$ mA are plotted for the different values of the externally applied electrostatic potential V_{dc}. Low voltages cause negligible shift and the solutions almost coincide with the solution for $V_{dc} = 0$. As the external potential increases, the electrostatic force acts as an extra restoring force, which corresponds to an additional stiffness causing the resonant

Out-of-Plane Resonant Capacitive Detection Magnetometer

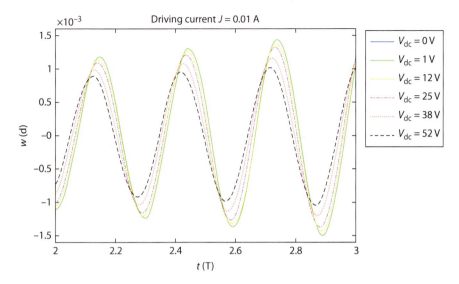

FIGURE 3.19
Solutions of the governing PDE for different V_{dc} values and a driving current amplitude of $J = 10$ mA.

frequency of the structure to shift higher. The maximum deflection of the beam at resonance also decreases slightly as the external tuning potential increases due to the added force.

Next, we calculate the resonant frequency to be expected when the sensor vibrates under the action of the Lorentz force in the presence of an external potential V_{dc}. The external potential is varied between 1 and 52 V. Numerical solutions are obtained for different values of the amplitude of the driving current. In Figure 3.20, the first resonant frequency of the sensor as a function of the externally applied electrostatic potential is plotted for a driving current with an amplitude between 2 and 70 mA. A linear fit is then applied to the model data in order to compare the results with those obtained from the experiment. For zero voltage, we get a constant equal to 19.2 kHz from the least square fit, which is close to the experimentally obtained value of 18.8 kHz.

The model accurately describes the tuning effect and the dependence on the driving current. The first resonant frequency shifts higher as the electrostatic potential between the comb fingers increases and the effect is much more pronounced when the amplitude of the driving current is low (<20 mA). This should be expected since for small driving currents, the amplitude of the Lorentz force is small and the resulting maximum deflection at resonance is also relatively small. In this regime of small vibrations, the exponential term due to the electrostatic attraction is significant and contributes to the total restoring force, thereby altering the resonant frequency of the structure. As the amplitude of the driving current increases, the maximum deflection at resonance also increases. This causes the beam to bend further

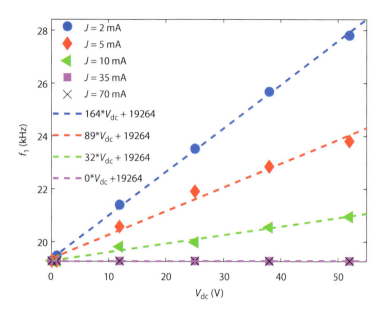

FIGURE 3.20
Frequency shifts estimated as a function of the external V_{dc} estimated from the model for different values of driving current amplitudes.

away from the substrate where the exponential term of the electrostatic attraction vanishes. The shift when the driving current is low is calculated quite accurately; however, for higher amplitudes (beyond 20 mA) the model gives a zero shift while our experiments suggest that even for 70 mA a small tuning effect should be expected. The main reason for this failure is probably because the model does not account for thermal stress effects due to joule heating or for nonlinear effects due to large deflections that might rise when the driving current is sufficiently high. Finally, an indirect consequence of the dependence of the tuning effect on the amplitude of the driving current is that it should also depend on the quality factor of the vibrating structure that, for a given stimulus, controls the maximum deflection at resonance. Thus, similar structures with different Q will behave differently when we attempt to tune their frequency of resonance using an external electrostatic field. For a given driving current, the device with the higher Q will deflect more and will be less prone to tuning. The model also extracts the modal shape of the beam and the maximum deflection at resonance, which is normally observed at the midpoint in the case of the first resonant frequency. In Figure 3.21, the modal shape for an excitation current with amplitude 10 mA is plotted for different values of the dc bias. The x-axis of the plot corresponds to the dimensionless length of the beam and the y-axis corresponds to the dimensionless (measured in units of d) out-of-plane deflection.

The out-of-plane deflection is maximum at the midpoint (a distance of almost 1 mm from the two anchors) as should be expected in the case of the

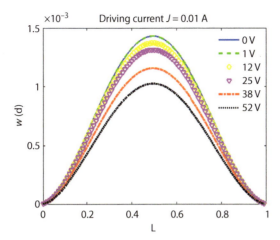

FIGURE 3.21
Modal shape of the beam when it vibrates at its first resonant frequency under the action of Lorentz force for a current with amplitude $J = 10$ mA.

first mode of resonance. The amplitude decreases with the tuning bias from 440 nm for $V_{dc} = 0$ V to 310 nm for $V_{dc} = 52$ V. It is interesting to note that if we double the amplitude of the driving current, the amplitude at resonance for $V_{dc} = 0$ V doubles to 880 nm as should be expected; however, the response for $V_{dc} = 52$ V is close to 850 nm, which is much larger than twice the value for $J = 10$ mA. As the excitation current increases, the amplitude of the Lorentz force also increases resulting in larger amplitudes at resonance. As the comb fingers on the vibrating beam move away from the stationary fingers at distances over 1.2 µm, the electrostatic force exerted on them (Equation 3.17 and Figure 3.18) decreases fast. For an excitation current $J = 70$ mA, the maximum deflection at resonance is 3.1 µm and does not change with V_{dc}. This is lower than the experimentally obtained value of 3.8 µm for $J = 48$ mA rms, which suggests that possibly thermal and nonlinear effects that were ignored in our analysis become dominant at these high currents. We finally obtain the phase diagram of Equation 3.21 in Figure 3.22, after adding a damping term, for $K = 0$, 1, and 5, assuming an excitation current with amplitude $J = 5$ mA. As previously discussed, constant K is a measure of the relative importance of the added stiffness due to the tuning effect over the mechanical stiffness of the structure. $K = 0$ corresponds to no tuning ($V_{dc} = 0$), $K = 1$ corresponds to tuning with a V_{dc}, which produces a restoring force with β equal to the equivalent stiffness of the beam (approximately 20 N/m), and so on.

3.5.3 Experimental Validation and Discussion

In order to check the validity of the model, the effect of dc bias on the resonant frequency of the beam for different values of the driving current was

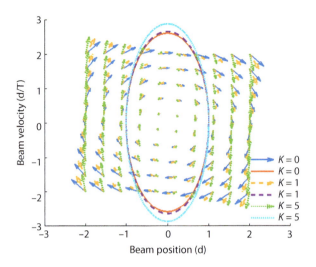

FIGURE 3.22
Phase diagram for different damping value ($K = 0$, 1, and 5), when the excitation current amplitude $J = 5$ mA.

investigated. The dc bias was varied between 0 and 52 V for two rms values of the driving current. The first series of measurements was obtained using a low rms current of $J = 2$ mA. In this case, both thermal stress and the amplitude of the vibration were expected to be low. The second series of measurements was obtained using a high driving current $J = 70$ mA. All measurements were performed at ambient temperature using a Polytec MSA-500 laser Doppler vibrometer. In Figure 3.23, experimentally obtained values of the first resonant frequency of the beam for dc bias between 0 and 52 V are plotted for the two different actuating currents. Black squares correspond to $J = 2$ mA, while red squares correspond to $J = 70$ mA [39].

Based on these measurements, we can confirm that the first resonant frequency increases with dc bias and shifts almost 38% when a dc voltage of 52 V is applied and a 2 mA rms current is flowing between the beam ends. Under these conditions, both the amplitude of vibration and the thermally induced stress due to joule heating effects are small. As a result, there is good agreement between our model predictions and the experimentally obtained results. To facilitate comparisons, we apply a linear fit (solid black line) to the experimentally obtained points and we notice that the resulting linear equation ($18{,}800 + 144 \times V_{dc}$) is very similar to the one obtained previously from the model ($19{,}264 + 164 \times V_{dc}$).

For the high intensity actuating current, the tuning effect is less pronounced and the frequency shifts about 7% higher. We see that the tuning effect is higher at lower displacements. This is obviously linked to the range of the electrostatic force, which is decreasing exponentially to zero as the out-of-plane deflection of the beam increases. For small displacements, the

Out-of-Plane Resonant Capacitive Detection Magnetometer

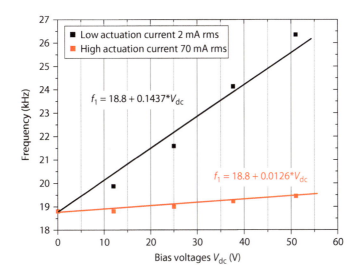

FIGURE 3.23
Measured resonant frequency as a function of the biasing voltage for low and high actuation currents.

electrostatic force is almost linear and affects both the beam displacement and the resonant frequency. For large deflections, however, the effect of the electrostatic force becomes small and, as a result, the frequency shift should diminish as well. In Figure 3.23, a linear fit and the corresponding analytic equation relating the resonant frequency to the bias voltages for the case of $J = 70$ mA rms are also presented. From this, we immediately see that our model fails to predict the shift under these conditions probably because the combined effect of thermal stress and large deflections cannot be neglected in this case.

We can deduce from Section 3.5.2 that the increase in resonant frequency is mainly due to the electrostatic force that arises between the movable and the stationary interdigitated electrodes when a dc bias is applied between them. This force is attractive and linear in case displacements are kept low, as illustrated by Figure 3.23. In this configuration, an additional electrostatic stiffness component, K_{elec}, is generated and added to the initial elastic stiffness, K_{elas}, of the clamped-clamped resonant beam. For larger magnitude displacements, the electrostatic force is no longer linear and the electric stiffness has less effect on the resonance frequency. We note that the frequency change in this structure was different from that reported in the literature, since the external electrostatic force is generated in the z-direction and not laterally [15,16]. Tuning frequencies with electrostatic actuation is very important and interesting because it can serve as a feedback mechanism to compensate for measurement errors, such as the nonlinear problem encountered due to the actuating current.

3.6 Conclusion

In this chapter, a MEMS magnetometer with capacitive detection, fabricated in a SOIMUMPS process, was presented and characterized. This magnetometer relies on a set of comb fingers for capacitive detection. A tuning method was applied to this kind of magnetometer by setting a dc voltage between the comb fingers that led to a change in the structure's equivalent spring constant. The experimental values show that the curve is nearly linear for specific bias voltages and the resonant frequency increases by 38% from its original value. This shift is very interesting especially when the displacements of the fingers are kept low since the electrostatic forces are then linear and provide an electrical stiffness component that affects the mechanical stiffness of the vibrating beam. Electrostatic tuning in this type of magnetometer can serve as a feedback mechanism to compensate for measurement errors, and may also increase the sensor performance by increasing the bandwidth of the measurement of an alternating magnetic field.

References

1. T.A. Roessig, R.T. Howe, A.P. Pisano and J.H. Smith, Surface micromachined resonant accelerometer, in *Proceedings of the International Conference on Solid-State Sensors and Actuators (Transducers'97)*, pp. 859–862, Chicago, 16–19 June 1997.
2. Y. Omura, Y. Nonomura and O. Tabata, New resonant accelerometer based on rigidity change, in *Proceedings of the International Conference on Solid-State Sensors and Actuators (Transducers'97)*, pp. 855–858, Chicago, 16–19 June 1997.
3. L. Lin, R.T. Howe and A.P. Pisano, Microelectromechanical filters for signal processing, *Journal of Microelectromechanical Systems*, Vol. 7, No. 3, pp. 286–294, 1998.
4. K. Wang and C.T.-C. Nguyen, High-order micromechanical electronic filters, in *Proceedings of the IEEE International Microelectromechanical Systems Workshop*, pp. 25–30, Nagoya, Japan, 26–30 January 1997.
5. C.T.-C. Nguyen, Frequency-selective MEMS for miniaturized low power communication devices, *IEEE Transactions on Microwave Theory Technology*, Vol. 47, No. 8, pp. 1486–1503, 1999.
6. Y. Oh, B. Lee, S. Baek, H. Kim, J. Kim, S. Kang and C. Song, Surface-micromachined tunable vibratory gyroscope, in *Proceedings of the IEEE International Microelectromechanical Systems Workshop*, pp. 272–277, Nagoya, Japan, 26–30 January 1997.
7. K. Tanaka, Y. Mochida, M. Sugimoto, K. Moriya, T. Hasegawa, K. Atsuchi and K. Ohwada, A micromachined vibrating gyroscope, *Sensors and Actuators*, Vol. 50, pp. 111–115, 1995.
8. S.K. Clark and K.D. Wise, Pressure sensitivity in anisotropically etched thin-diaphragm pressure sensors, *IEEE Transactions on Electron Devices*, Vol. 26, No. 12, pp. 1887–1896, December 1979.

9. A.L. Herrera-May, P.J. García-Ramírez, L.A. Aguilera-Cortés, J. Martínez-Castillo, A. Sauceda-Carvajal, L. García-González and E. Figueras-Costa, A resonant magnetic field microsensor with high quality factor at atmospheric pressure, *Journal of Micromechanics and Microengineering*, Vol. 19, No. 1, pp. 1–11, January 2009.
10. E. Manjarrez, A.L. Herrera-May, L.A. Aguilera-Cortés and P.J. García-Ramírez, Resonant magnetic field sensors based on MEMS technology, *Sensors Journal*, Vol. 39, No. 9, pp. 7785–7813, 2009.
11. M.T. Todaro, L. Sileo and M. De Vittorio, Magnetic field sensors based on microelectromechanical systems (MEMS) technology, in *Magnetic Sensors: Principles and Applications*, Chapter 6, K. Kuang Ed., Intech, Croatia, 2012.
12. V. Kaajakari, *Practical MEMS: Design of Microsystems, Accelerometers, Gyroscopes, RF MEMS, Optical MEMS, and Microfluidic Systems*, Small Gear Publishing, Las Vegas, NV, 2009.
13. S. Brugger and O. Paul, Field-concentrator-based resonant magnetic sensor with integrated planar coils, *Journal of Microelectromechanical Systems*, Vol. 18, No. 6, pp. 1432–1443, December 2009.
14. H. Emmerich and M. Schofthaler, Magnetic field measurements with a novel surface micromachined magnetic-field sensor, *IEEE Transactions on Electron Devices*, Vol. 47, pp. 972–977, 2000.
15. J.W. Kang, H. Guckel and Y. Ahn, Amplitude detecting micromechanical resonating beam magnetometer, in *The Eleventh International Workshop on Micro Electro Mechanical Systems*, pp. 372–377, Heidelberg, Germany, 1998.
16. M. Offenberg, F. Larmer, B. Elsner, H. Munzel and W. Riethmuller, Novel process for a monolithic integrated accelerometer, in *Eighth International Conference on Solid-State Sensors and Actuators (Transducers'95)*, pp. 582–592, Stockholm, Sweden, 1995.
17. M. Offenberg, H. Münzel, D. Schubert, O. Schatz, F. Lärmer, E. Müller, B. Maihöfer and J. Marek, Acceleration sensor in surface micromachining for air-bag applications with high signal/noise ratio, in *Proceedings of the International Congress and Exposition*, pp. 35–41, Detroit, MI, 26–29 February 1996.
18. D. Ullmann et al., Sensors for automotive safety systems in surface micromachining, VDI Rep. 1415, 1998.
19. B. Bahreyni and C. Shafai, A resonant micromachined magnetic field sensor, *IEEE Sensors Journal*, Vol. 7, No. 9, pp. 1326–1334, 2007.
20. M.J. Thompson and D.A. Horsley, Resonant MEMS magnetometer with capacitive read-out, in *2009 IEEE Sensors Conference*, pp. 992–995, Christchurch, New Zealand, 25–28 October 2009.
21. Z. Kádár, A. Bossche, P.M. Sarro and J.R. Mollinger, Magnetic-field measurements using an integrated resonant magnetic-field sensor, *Sensors Actuators A Physical*, Vol. 70, No. 3, pp. 225–232, 1998.
22. J. Tucker, D. Wesoleck and D. Wickenden, An integrated CMOS MEMS xylophone magnetometer with capacitive sense electronics, *2000 NanoTech*, pp. 1–5, Houston, TX, 2002.
23. J. Kyynäräinen et al., A 3D micromechanical compass, *Sensors and Actuators A*, Vol. 142, No. 2, pp. 561–568, 2008.
24. D. Ren, L. Wu, M. Yan, M. Cui, Z. You and M. Hu, Design and analyses of a MEMS based resonant magnetometer, *Sensors Journal*, Vol. 9, No. 9, pp. 6951–6966, 2009.

25. A.L. Herrera-May, L.A. Aguilera-Cortés, P.J. García-Ramírez, N.B. Mota-Carrillo, W.Y. Padrón-Hernández and E. Figueras, Development of resonant magnetic field microsensors: Challenges and future applications, in *Microsensors*, I. Minin (Ed.), Chapter 3, InTech, Rijeka, Croatia, 2010.
26. P. Gkotsis, M. Lara-Castro, F. López-Huerta, A.L. Herrera-May and J.-P. Raskin, Mechanical characterization and modeling of Lorentz force based MEMS magnetic field sensors, *Solid-State Electronics*, Vol. 112, pp. 68–77, 2015.
27. J.A. Tapia, A.L. Herrera-May, P.J. García-Ramírez, J. Martínez-Castillo, E. Figueras, A. Flores and E. Manjarrez, Sensing magnetic flux density of artificial neurons with a MEMS device, *Biomedical Microdevices*, Vol. 13, No. 2, pp. 303–313, April 2011.
28. S. Ranvier, V. Rochus, S. Druart, H. Lamy, P. Rochus and L.A. Francis, Detection methods for MEMS-based xylophone bar magnetometer for pico satellites, *Journal of Mechanics Engineering and Automation*, Vol. 1, pp. 342–350, 2011.
29. M.H. Said, F. Tounsi, P. Gkotsis, M. Masmoudi and L.A Francis, MEMS-based clamped-clamped beam resonator capacitive magnetometer, *Transactions on Systems, Signals and Devices*, Vol. 9, No. 4, pp. 483–499, 2013.
30. W. Weaver, S.P. Timoshenko and D.H. Young, *Vibration Problems in Engineering*, 5th Edn, Wiley, New York, 1990.
31. M.H. Said, F. Tounsi, P. Gkotsis, M. Masmoudi and L.A Francis, A MEMS resonant magnetometer based on capacitive detection, *10th IEEE International Multi-Conference on Systems, Signals and Devices (SSD)*, Tunisia, 2013.
32. W.S. Ibrahim and W.G. Ali, A review on frequency tuning methods for piezoelectric energy harvesting systems, *Journal of Renewable Sustainable Energy*, Vol. 4, No. 6, pp. 29, November 2012.
33. K.B. Lee, L. Len and Y.H. Cho, A closed-form approach for frequency tunable comb resonators with curved finger contour, *Sensors and Actuators A*, Vol. 141, No. 2, pp. 523–529, 2008.
34. R. Todd and W. Liwei, Active frequency tuning for micro resonator by localized thermal stressing effects, *Sensors and Actuators A*, Vol. 91, No. 3, pp. 326–332, 2001.
35. G. Shim, R. Mimoto, S. Kumagai and M. Sasaki, Macro model study for nonlinear spring of tense torsion bar in gap-closing type electrostatic micromirror, *Journal of Mechanics Engineering and Automation*, Vol. 2, No. 7, pp. 446–454, 2012.
36. W. Sul Lee, K.C. Kwon, B.K. Kim, J.H. Cho and S.K. Youn, Frequency-shifting analysis of electrostatically tunable micro-mechanical actuator, *Journal of Modeling and Simulation of Microsystems*, Vol. 2, No. 1, pp. 83–88, 2001.
37. J.A. Pelesko and D.H. Bernstein, *Modeling MEMS and NEMS*, CRC Press, Boca Raton, FL, 2002.
38. S. Margulies, Force on a dielectric slab inserted in a parallel plate capacitor, *American Journal of Physics*, Vol. 52, No. 6, pp. 515–518, June 1984.
39. M.H. Said, F. Tounsi, P. Gkotsis, B. Mezghani and L.A Francis, A resonant microstructure tunability analysis for an out-of-plane capacitive detection MEMS magnetometer, *Microsystem Technologies*, July 2016, DOI:10.1007/s00542-016-3093-y.

4

Micromachined Inductive Contactless Suspension: Technology and Modeling

**Kirill V. Poletkin, Vlad Badilita, Zhiqiu Lu,
Ulrike Wallrabe, and Christopher Shearwood**

CONTENTS

4.1 Introduction 102
    4.1.1 Micromachined Electromagnetic Contactless Suspensions 105
    4.1.2 Micromachined Inductive Suspension 106
4.2 Technology 107
    4.2.1 Planar Coils Technology 107
    4.2.2 3-D Microcoil Technology 108
4.3 Modeling 110
    4.3.1 Analytical Model 112
        4.3.1.1 Static Behavior of MIS 115
        4.3.1.2 MIS Linear Model 117
        4.3.1.3 Stable Levitation 119
    4.3.2 Analysis of MIS Designs 119
        4.3.2.1 Preliminary Analysis of Stability of MIS Based on
                Plane Coils 121
        4.3.2.2 Analysis of MIS Design Based on Two 3-D Coils 123
Acknowledgments 128
References 128

ABSTRACT Microelectromechanical systems (MEMS) have been used in an ever-increasing number of applications, on the one hand, replacing traditional sensors based on the bulk technologies; and on the other hand, applied principally to new applications, for instance, in healthcare, energy harvesting, active safety systems, computer hard drives, mobile devices, and adaptive optics. One of the main features of MEMS technology is miniaturization, which offers a multitude of advantages. However, as a result of shrinking size, the sensitivity of a microsensor is dramatically decreased due to the scaling effect and the domination of friction over inertial forces in the microworld. In turn, these facts also limit the performance of a microactuator. An indisputable solution is levitation, which eliminates completely physical attachment and, consequently reduces drastically mechanical friction. In particular, a micromachined inductive contactless suspension as a promising technology, to realize mentioned above scenario, is discussed in this chapter.

4.1 Introduction

Microelectromechanical systems (MEMS) are used in an ever-increasing number of applications. On the one hand, they are replacing traditional sensors based on the bulk technologies and, on the other hand, they are applied principally to new applications, for instance, in health care, energy harvesting, active safety systems, computer hard disk drives, mobile devices, and adaptive optics. Due to the remarkable progress in microsensors and actuators, micromotors with a diameter of one-tenth of a millimeter [1,2], inertial microsensors having a size less than 1×1 mm^2 [3,4], or microgenerators with a diameter of a few millimeters [5] are possible. These devices have also confirmed the feasibility of other advanced device concepts for the further improvement of performances in MEMS.

One of the main features of MEMS technology is miniaturization, which offers a multitude of advantages [6]. Indeed, micromachined inertial sensors can be easily incorporated into human clothes (i.e., shoes) to provide autonomous navigation parameters compared with bulk inertial sensors. A microsensor can be comfortably and gently embedded into the human body to continuously monitor its medical parameters. Micro-objects can be much more cautiously handled and precisely transported by a micromanipulator and a microtransporter, respectively, than their macro-counterparts [7]. In fact, the Internet of everything and wearable interfaces are only made possible by the availability of MEMS technology.

However, as a result of its shrinking size, the sensitivity of a microsensor is dramatically decreased due to the scaling effect and the domination of the friction over inertial forces in the microworld. In turn, these facts also limit the performance of a microactuator. An indisputable solution is levitation, which completely eliminates the mechanical attachment, and consequently the friction, between the stationary and moving parts of a microsensor and a microactuator. Besides, for a microactuator, levitation provides the extension of a motion range of its moving parts and reduces dissipated or wasted energy significantly. Although different physical mechanisms can be used to provide levitation, such as a jet of gas, intense sound waves, or beams of laser light [8], only levitation by means of electric and magnetic fields or electromagnetic levitation are practicable and have already attracted a great deal of attention from researchers, giving a rise to a new generation of microsensors and microactuators (see Table 4.1).

TABLE 4.1
Micromachined Electromagnetic Contactless Suspensions

Types of Suspension	Sources of Force Field	Materials of Levitated Mass/Rotor	Applications	Technologies	Ref.
	Active	**Contactless Suspensions**			
Electric	Electrical field	Conductive material	Suspension	Surface micromachining	[9]
		Dielectric	Suspension	Bulk	[10]
		Semiconductors and conductive material	Accelerometer	Exotic ball micromachining	[11,12]
Magnetic	Direct current	Magnet	Gyroscope multisensor	Bulk micromachining	[13,14,16,17]
			Actuator	Surface micromachining	[18]
	Passive	**Contactless Suspensions**			
Inductive	Alternating current	Conductive material	Gyroscope	Surface micromachining	[19–21]
			Accelerator		[22,23]
			Suspension		[24]
				3-D microcoil	[25]
Diamagnetic	Magnet	Diamagnetic	Accelerometer	Mix of surface micromachining and bulk	[26]
			Motor	Surface micromachining	[27]
			Gyroscope	Bulk micromachining	[15]

(Continued)

TABLE 4.1 (CONTINUED)
Micromachined Electromagnetic Contactless Suspensions

Types of Suspension	Sources of Force Field	Materials of Levitated Mass/Rotor	Applications	Technologies	Ref.
Superconducting	Magnet	Superconductor	Actuator	Surface micromachining	[28]
			Suspension		[29]
Hybrid Suspensions					
Diamagnetic-Inductive	Magnet ac	Diamagnetic conductive material	Gyroscope	Mix of surface micromachining and bulk	[30]
Electrostatic-Inductive	Electrostatic field ac	Conductive material	Suspension	Mix of surface micromachining and 3-D microcoil	[31,43]

4.1.1 Micromachined Electromagnetic Contactless Suspensions

The main issue of employing levitation in MEMS relates to stability or stable levitation. Stable levitation (suspension) means that a levitated mass is restored to rest when it is slightly displaced in any linear and angular direction from an equilibrium position. According to Earnshaw's theorem, in a system of electric charges or magnetic monopoles, stable levitation is impossible. In other words, from the point of view of the Lagrange–Maxwell equations, the potential energy (stored energy) of such a system does not have a local minimum. Therefore, stable levitation in a micromachined suspension based on an electrostatic field is only reached by using active control, in which an electric field is varied in such a way that the levitated mass is held at an equilibrium position. Such an approach to stable levitation is called *active contactless suspension* (AS). However, a static magnetic field created by a magnet can itself imitate the control system if the levitated mass is, for instance, diamagnetic. Hence, the restrictions of Earnshaw's theorem are overcome and stable levitation in a static magnetic field becomes possible. This is referred to as *passive contactless suspension* (PS). A summary of achievements in the fabrication and application of micromachined active and passive contactless suspensions is shown in Table 4.1.

The use of either AS or PS has its own advantages as well as disadvantages. Nevertheless, the principal difference between AS and PS is that AS requires displacement sensors to control all motion of the levitated mass, but PS isn't constrained in the same way. Hence, in assuming other conditions being equal, the realization of a micromachined contactless suspension based on PS is easier than that based on AS. As seen from Table 4.1, there are three possible techniques to build a micromachined PS, namely, diamagnetic, superconducting, or inductive contactless suspension. A key advantage of the first technique is that a permanent magnet is used. Due to this fact, we have a manifest benefit, which is stable levitation without the requirement of direct energy consumption [32]. Unfortunately, the load bearing capability of a diamagnetic suspension is small, which is reflected in its current performances and applications. The second technique requires a cryogenic environment, which becomes the major limit for the application of a superconducting suspension. An inductive suspension does not share the disadvantages previously mentioned. Moreover, recent achievements in the development of a micromachined inductive suspension (MIS), which established its micromachining fabrication process [35] and led to dramatically reducing heat dissipation [45], announce it as a very promising candidate as an integrated element for levitating microsystems.

Now, consider for a moment, the benefits from combining a diamagnetic and inductive suspension. Such a hybrid suspension is possible by the fabrication of the levitated mass as a multilayer consisting of conductive and diamagnetic materials, and for levitation, two sources of force fields are required, namely, an alternating current (ac) feeding the coils and a permanent magnet. As a result, a micromachined PS is created, in which, on the one hand, using an inductive suspension increases the carrying capability of the diamagnetic suspension and, on the other hand, the inductive suspension dissipates less energy because of the diamagnetic contribution. As seen, this hybrid suspension improves the energy performance of the PS in comparison with that of the diamagnetic or inductive suspension separately. It is worth noting that the hybrid suspension based on diamagnetic and inductive levitation was successfully fabricated and demonstrated in a micromachined gyroscope application by Liu et al. in [30]. Another hybrid suspension based on electrostatic actuation and induction levitation has recently been reported in [31]. This hybrid suspension is capable of adjusting dynamics by means of changing the values of its stiffness components and operating as a bistable microactuator. Moreover, such a hybrid suspension provides a platform to obtain a quasi-zero spring constant as was proposed in [33]. A micromachined prototype of this hybrid suspension and preliminary results of its experimental investigations accompanied by quasi-exact nonlinear modeling are presented in [43]. Hence, a new dynamic performance in the micromachined PS can be demonstrated by means of this electrostatic-inductive hybrid suspension. Thus, the hybrid suspension becomes a very promising technique to provide further improvements in the performance of a micromachined contactless suspension and, consequently, the expansion of its application.

4.1.2 Micromachined Inductive Suspension

The MIS is one of the key techniques to realizing a micromachined PS as well as a hybrid suspension, as shown in Table 4.1. As a rule, the MIS consists of a coil and a conductive proof mass (PM). The coil is fed by a high-frequency ac and generates a time-variable magnetic flux in space, which in turn induces an eddy current within the PM. Due to Ampere's law, the interaction between the eddy and the coil current produces a repulsive force between the coil and the PM. The repulsive force is the cause of the PM levitation.

As an illustrative example, consider the design of the MIS shown in Figure 4.1, which provides stable levitation of the disk-shaped PM. Originally, this design was proposed and a proof of concept demonstrated in [24]. Although the operating principle is the same as just mentioned, to provide stable levitation two coils are required, namely, a stabilization and

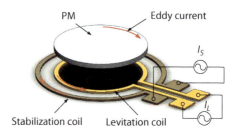

FIGURE 4.1
Design of MIS for stable levitation.

a levitation coil. Both coils are fed by ac at a high frequency, but with a 180 phase shift between the stabilization and levitation coil. Due to its intrinsic simplicity, this design is widely used in other prototypes of MIS reported in the literature. Besides, a simple extension of this design can provide phased poles to rotate the PM, as was demonstrated by Shearwood [34]. It is obvious that in depending on the MIS application, there are other MIS designs (see Table 4.1); however, the common point for all of them encompasses two main factors: the microcoils fabrication and stable levitation of the PM.

This chapter aims to consider these two issues in detail. Namely, Section 4.2 considers micromachined technologies, which have already been established for microcoil fabrication. These technologies are surface micromachining and 3-D coil technology. Section 4.3 discusses analytical mathematical methods, which have been extensively applied to model the MIS stability, dynamics, and statics.

4.2 Technology

The MIS has been actively studied since 1995, when Shearwood [24] fabricated the first prototype targeted at applications including a micromachined motor and gyroscope. Then, in 2006, the Zhang group presented the MIS with an improved planar coil design [20]. An alternative coil design in the shape of a rectangular spiral, which also provides stable levitation, was employed in the microgyroscope prototype reported in [21]. All the aforementioned research groups used surface micromachining technology for coil fabrication. Due to this technology, the coils and levitated PM dissipated a lot of energy. As a result, the observed temperatures of the coils and PM were as high as 600°C and 100°C, respectively. In 2011, using recently developed 3-D coil technology [35], the Badilita group dramatically reduced the energy dissipation in the MIS [25] and proposed a new application for the MIS as a micromachined transporter. This group also established the technological direction for further improvement of the MIS.

4.2.1 Planar Coils Technology

In general, microcoil fabrication for the MIS focuses on the following target characteristics: a high electrical quality factor and hence low electrical resistance, an increase in the number of winding turns to reduce the amplitude of the coil current, electrical winding insulation, and low manufacturing and assembly costs.

Designs of the planar microcoils used in the MIS are shown in Figure 4.2. Both designs of the MIS shown in Figures 4.1 and 4.2a provide only stable levitation of the disk-shaped PM, while Figure 4.2b and c demonstrate designs that in addition to stable levitation allow us to rotate the PM. The rotation is achieved by using poles, which are sequentially excited by a high-frequency current. The frequency of the excitation of the poles defines the PM rotation speed.

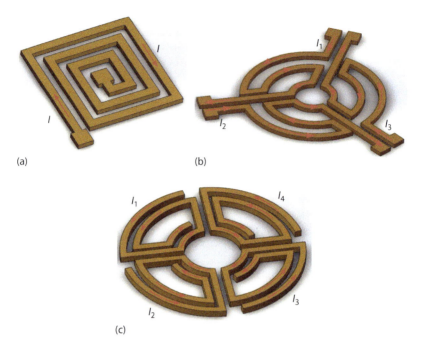

FIGURE 4.2
Designs of planar microcoils: I is the electric current.

Planar microcoil fabrication exploits surface macromachining technology, which includes the following essential techniques: metal evaporation, electroplating, and wet chemical etching. An outline of the fabrication process for a one-layer planar microcoil is shown in Figure 4.3.

As seen from Figure 4.3, the fabrication of a one-layer coil is relatively simple. As a result, within the framework of one layer, designing the coil becomes flexible allowing for the design of a complex coil structure. However, the fabrication complexity is dramatically increased once the number of layers and, consequently, the number of coil turns are increased. Moreover, on the one hand, the need to increase coil turns arises due to the technological problem of obtaining a low electrical resistance coil, which limits the current amplitude in the coil and, as a result, the performance of the MIS. On the other hand, there is a physical limit to the maximum possible current density of metal.

4.2.2 3-D Microcoil Technology

Using 3-D microcoil technology, a microcoil with an unlimited number of turns can be fabricated. Moreover, irrespective of the number of turns, only two masks are required for the fabrication. The typical design of 3-D coils, which is applied to the MIS (3-D MIS), is shown in Figure 4.4a. This design includes two 3-D microcoils, namely, stabilization and levitation, which have

Micromachined Inductive Contactless Suspension

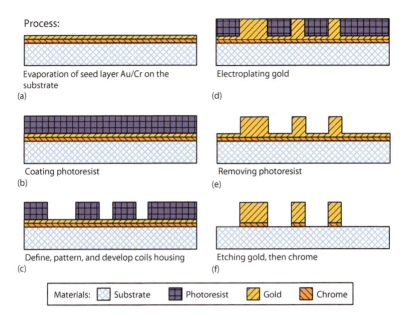

FIGURE 4.3
Outline of the fabrication process for a one-layer planar microcoil.

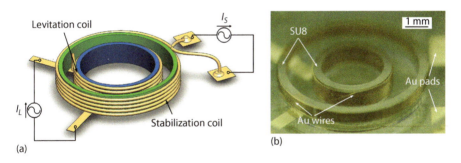

FIGURE 4.4
(a) Design of 3-D microcoils for the MIS; (b) fabricated 3-D microcoils.

the same targets as the coils in the planar design shown in Figure 4.1. As an example, Figure 4.4b shows the fabricated 3-D microcoil structure according to the design of two 3-D microcoils shown in Figure 4.4a.

The fabrication process consists of three main steps. In the first step, using surface micromachining, pads for electrical contacts are fabricated on the substrate by the same process outlined in Figure 4.3, where the first mask is used. In the second step, usually 600–700 μm-thick SU-8 2150 is cast on the substrate. Using the second mask, the cylindrical pillars are structured by ultraviolet (UV) lithography. An outline of this process is shown in Figure 4.5. In the last step, the coils are manufactured using an automatic wirebonder, which allows us to freely define the total number of windings

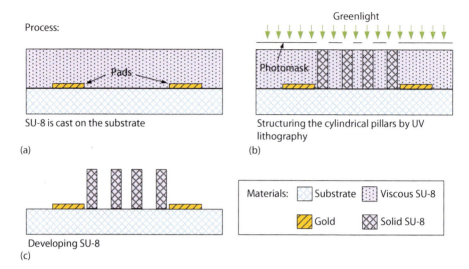

FIGURE 4.5
Outline of the fabrication process of cylindrical pillars.

per coil, the pitch between the windings, and the number of winding layers. Although using a wirebonder for coil fabrication is a serial process, this process is very fast—hundreds of milliseconds per coil depending on the exact dimensions—and perfectly integrates with traditional MEMS processes [35].

Thus, 3-D microcoils as opposed to planar microcoils dramatically increase the number of winding turns and, at the same time, drastically reduce the fabrication complexity. By embracing 3-D microcoils, energy dissipation is significantly reduced. The measured temperatures of coils and levitated PM were around 100°C and 55°C, respectively, for a PM levitation height of 120 μm, and the levitation coil with 20 windings and the stabilization coil with 15 windings as reported in [36]. The abovementioned advantages of 3-D microcoils, which are supported by using 3-D coil technology, further improve the MIS performance and open a new application of MIS, for instance, in microtransporters. Such a 3-D structure, which was fabricated in cooperation with the Laboratory for Microactuators and the Laboratory for Simulation at IMTEK, Freiburg University, and can be applied as a microtransporter, is shown in Figure 4.6. Figure 4.6 also shows that 3-D coil technology allows for the fabrication of noncylindrical coils.

4.3 Modeling

As previously mentioned, upon designing the MIS a key issue is to provide stable levitation of PM. This means that a proposed design of MIS, namely, geometrical parameters of coils and PM, and the values of currents in each

Micromachined Inductive Contactless Suspension 111

FIGURE 4.6
Wirebonded nonsolenoidal structure realized in cooperation with the Laboratory for Microactuators and the Laboratory for Simulation.

coil, must fulfill the condition for stable levitation. Considering the MIS as an electromechanical system, this condition for stable levitation results from the existence of a local minimum of potential energy (stored energy) of such a system. Thus, in order to obtain this condition for stable levitation for a particular design of the MIS, its potential energy should be known and presented in analytical form. In general, the MIS is a system with distributed parameters and is described by a set of partial differential equations. Hence, the presentation of the MIS potential energy in analytical form has an obvious difficulty. Due to this fact, the first design of the MIS (see Figure 4.1), which provided stable levitation of the disk-shaped PM, was a result of pure experiential development of the Shearwood group.

However, for the case when the MIS consists of a disk-shaped PM and a ring-shaped coil fed by a high-frequency current, I, as shown in Figure 4.7a, the distribution of the eddy current density within the PM induced by the

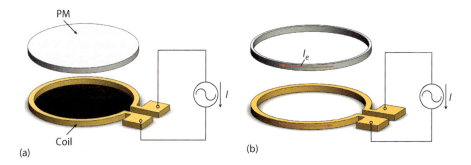

FIGURE 4.7
(a) Design of the MIS with the disk-shaped PM and ring-shaped coil; (b) idealization: I_e is the eddy current.

time-variable magnetic flux is well known. The maximum density of the eddy current is distributed along the edge of the PM. The value of the eddy current density is exponentially reduced from the PM circumference to its center. Hence, it can be assumed that the eddy current flowing along the edge of the PM and the coil current define the acting forces between the disk-shaped PM and the ring-shaped coil; the design shown in Figure 4.7a can be idealized and represented as the lumped system depicted in Figure 4.7b. The interaction between the coil and the disk can be considered as the interaction between two ring-shaped coils. Now, in the framework of this lumped system, the analytical form of the potential energy can be written and then applied to a more complex design consisting of more coils. Moreover, knowing the MIS potential energy, the dynamic and static characteristics of the MIS can be evaluated. Thus, the analytical model of the MIS can be obtained, which describes the MIS behavior qualitatively. It becomes a powerful mathematical tool for MIS modeling that can be used to develop new MIS designs with, for instance, an improved dynamic performance [37–39,44].

4.3.1 Analytical Model

To describe the behavior of the PM with respect to the coil shown in Figure 4.7a, let us assign the following coordinate frames (CF), namely, the fixed CF XYZ to the coil and the movable CF xyz to the PM. The Z-axis of the fixed CF is coincident with the coil axis, which is passed through the coil center and perpendicular to the coil plane. The origin A of the fixed CF is located at the intersection of the coil axis and its equatorial plane. Meanwhile, the axes of the movable CF coincide with the principal axes of inertia of the disk-shaped PM.

An arbitrary position of the PM, when the PM is disturbed, is shown in Figure 4.8. Assume that the origin B of the movable CF is coincident with the center of mass of PM and, in rest (equilibrium state), with the origin O. The position of the origin O in space with respect to the fixed CF is characterized by a vector $\vec{r}_o = (X_o, Y_o, Z_o)$. In particular, the vector $\vec{r}_o$ can be defined as $\vec{r}_o = (0,0,h)$, where h is the levitation height of the PM. Also, let us introduce the following auxiliary CFs, namely, $x_1y_1z_1$, the origin of which is coincident with point O, $x_2y_2z_2$, and $x_3y_3z_3$, the origins of which are assigned to point B. In addition to the axes of CF $x_1y_1z_1$, CF $x_2y_2z_2$ and fixed CF are collinear. The position of CF $x_3y_3z_3$ with respect to CF $x_2y_2z_2$ is defined by the angle α, which characterizes the angular misalignment of CF $x_3y_3z_3$ relative to the Bz_2 axis. Besides, the Bx_3 is coincident with the x-axis of the movable CF. In turn, the movable CF has angular misalignment with respect to CF $x_3y_3z_3$ defined by the angle φ relative to the Bz_3 axis.

Due to the particular design of the MIS under consideration, and since the linear displacements of PM are assumed to be small, the mechanical properties of the suspension can be considered isotropic in the radial direction and independent of the angle α. Also, it is assumed that there is no rotation of the

Micromachined Inductive Contactless Suspension

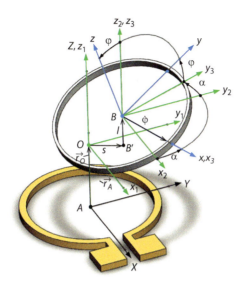

FIGURE 4.8
An arbitrary position of the PM, with respect to the fixed CF: $x_1y_1z_1$, $x_2y_2z_2$, and $x_3y_3z_3$ are auxiliary CFs (green color); $\vec{r}_A$ and $\vec{r}_B$ are vectors characterized by the position of point B with respect to the fixed and auxiliary $x_1y_1z_1$ CFs, respectively; $s = \sqrt{(x_1^2 + y_1^2)}$, l and φ are the generalized coordinates of the mechanical part of the MIS.

PM relative to the Bz axis. Then, the generalized coordinates for describing the mechanical part of the MIS can be introduced as follows. The generalized coordinate, $s = \sqrt{(x_1^2 + y_1^2)}$, characterizes the linear displacement of the center of mass of the PM in the radial direction. The generalized coordinate, l, characterizes the linear displacement of the center of mass of the PM along the Oz_1, and the generalized coordinate, φ, characterizes the angular displacement of the PM relative to an axis coincident with the principal axis of inertia of the disk-shaped PM lying on its equatorial plane, as shown in Figure 4.8. It is worth noting that the introduced generalized coordinates allow us to reduce the system from six to three degrees of freedom to describe the mechanical part of the MIS.

The electrical part of the MIS can be represented by two electrical circuits having mutual induction between the coil and PM, L_m. The first circuit contains the inductance of the coil, L_c, and its resistance, R_c, and the source of ac, i_c, while the second circuit has only the inductance of the PM, L_{pm}, and its resistance, R_{pm}. The ac can be described by $i_c = Ie^{j\omega t}$, where I is the amplitude, ω is the frequency, and $j = \sqrt{(-1)}$ is an imaginary unit. Due to the mutual inductance, L_m, between the coil and PM, the eddy current, i_e, is induced. Hence, the current, i_e, is taken as the velocity of the generalized coordinate of the electrical part of the MIS. Finally, the state of the MIS as an electromechanical system can be described by four generalized coordinates.

Adopting the foregoing generalized coordinates and assumptions, the model of the MIS can be written using the Lagrange–Maxwell equations as follows:

$$\begin{cases} \dfrac{d}{dt}\left(\dfrac{\partial L}{\partial \dot{i}_e}\right) + \dfrac{\partial \Psi}{\partial \dot{i}_e} = 0; \\ \dfrac{d}{dt}\left(\dfrac{\partial L}{\partial \dot{l}}\right) - \dfrac{\partial L}{\partial l} + \dfrac{\partial \Psi}{\partial \dot{l}} = F_l; \\ \dfrac{d}{dt}\left(\dfrac{\partial L}{\partial \dot{s}}\right) - \dfrac{\partial L}{\partial s} + \dfrac{\partial \Psi}{\partial \dot{s}} = F_s; \\ \dfrac{d}{dt}\left(\dfrac{\partial L}{\partial \dot{\varphi}}\right) - \dfrac{\partial L}{\partial \varphi} + \dfrac{\partial \Psi}{\partial \dot{\varphi}} = M_\varphi, \end{cases} \quad (4.1)$$

where:
$L = T(\dot{l},\dot{s},\dot{\varphi}) - \Pi(l) + W_m(l,s,\varphi,i_c,i_e)$ is the Lagrange function for the PM–coil system
$T(\dot{l},\dot{s},\dot{\varphi})$ and $\Pi(l)$ are the kinetic and potential energies of the system
$W_m(l,s,\varphi,i_c,i_e)$ is the energy stored in the electromagnetic field
$\Psi(\dot{l},\dot{s},\dot{\varphi},i_c,i_e)$ is the dissipation function of the system
F_l, F_s, and M_φ are the generalized forces and torque acting on the PM relative to the appropriate generalized coordinates l, s, and φ, respectively

Assuming that the acceleration due to gravity is directed along the Z-axis, the potential energy Π is dependent on the generalized coordinate l only.
The kinetic energy is

$$T = \frac{1}{2}m\dot{l}^2 + \frac{1}{2}m\dot{s}^2 + \frac{1}{2}J\dot{\varphi}^2, \quad (4.2)$$

where:
m is the mass of PM
J is the moment of inertia of the PM about the axis lying on its equatorial plane

The equations for potential energy and the energy stored within the electromagnetic field are

$$\Pi = mgl, \quad (4.3)$$

$$W_m = \frac{1}{2}L_c i_c^2 + L_m(l,s,\varphi)i_c i_e + \frac{1}{2}L_{pm}i_e^2, \quad (4.4)$$

respectively. Note that in Equation 4.4, the mutual inductance L_m is a function of the generalized coordinates l, s, and φ. Thus, the Lagrange function for the system under consideration can be written as

$$L = \frac{1}{2}m\dot{l}^2 + \frac{1}{2}m\dot{s}^2 + \frac{1}{2}J\dot{\varphi}^2 - mgl$$
$$+ \frac{1}{2}L_c i_c^2 + L_m(l,s,\varphi)i_c i_e + \frac{1}{2}L_{pm}i_e^2. \quad (4.5)$$

The dissipation function is

$$\Psi = \frac{1}{2}\mu_l \dot{l}^2 + \frac{1}{2}\mu_s \dot{s}^2 + \frac{1}{2}\mu_\varphi \dot{\varphi}^2 + \frac{1}{2}R_c i_c^2 + \frac{1}{2}R_{pm}i_e^2, \quad (4.6)$$

where μ_l, μ_s, and μ_φ are the damping coefficients.

In this study, the coefficients of dissipative function are assumed to be constant. Substituting Equations 4.5 and 4.6 into set 4.1, we obtain the following model of the MIS:

$$\begin{cases} L_{pm}\dfrac{di_e}{dt} + L_m \dfrac{di_c}{dt} \\ \quad + \left(\dfrac{\partial L_m}{\partial l}\dot{l} + \dfrac{\partial L_m}{\partial s}\dot{s} + \dfrac{\partial L_m}{\partial \varphi}\dot{\varphi}\right)i_c + R_{pm}i_e = 0; \\ m\ddot{l} + \mu_l \dot{l} + mg - \dfrac{\partial L_m}{\partial l}i_c i_e = F_l; \\ m\ddot{s} + \mu_s \dot{s} - \dfrac{\partial L_m}{\partial s}i_c i_e = F_s; \\ J\ddot{\varphi} + \mu_\varphi \dot{\varphi} - \dfrac{\partial L_m}{\partial \varphi}i_c i_e = M_\varphi, \end{cases} \quad (4.7)$$

Equation 4.7 is a set of nonlinear equations describing the dynamics of the levitated disk-shaped PM, suspended by the MIS in space.

4.3.1.1 Static Behavior of MIS

In the framework of developing the linear model of the MIS, the function of mutual inductance L_m, which is a function of the generalized coordinates, can be represented in quadratic form as follows:

$$L_m(l,s,\varphi) = c_0 + c_l l + c_s s + c_\varphi \varphi$$
$$+ \frac{1}{2}c_{ll}l^2 + \frac{1}{2}c_{ss}s^2 + \frac{1}{2}c_{\varphi\varphi}\varphi^2 + c_{ls}ls + c_{l\varphi}l\varphi + c_{s\varphi}s\varphi, \quad (4.8)$$

where the coefficients c_0, c_{nm} ($n = l,s,\varphi$; $m = l,s,\varphi$) can be defined as follows. Assuming that the equilibrium position of the PM is occupied at the point O with the coordinates $l = h$, $s = 0$, and $\varphi = 0$ (these coordinates of the equilibrium point of the PM follow directly from the experimental observation of MIS [20,25,34,38]). Then, we can write

$$c_0 = L_m(h,0,0), \tag{4.9}$$

$$c_n = \left.\frac{\partial L_m}{\partial n}\right|_{\substack{l=h\\s=0\\\varphi=0}}, \ (n = l,s,\varphi), \tag{4.10}$$

$$c_{nm} = \left.\frac{\partial^2 L_m}{\partial n \partial m}\right|_{\substack{l=h\\s=0\\\varphi=0}}, \ (n = l,s,\varphi), (m = l,s,\varphi). \tag{4.11}$$

In the static state, the influence of velocities $\dot{l},\dot{s},\dot{\varphi}$, accelerations $\ddot{l},\ddot{s},\ddot{\varphi}$, and the generalized forces and torque F_l, F_s, M_φ acting on the PM are neglected. Hence, set 4.7 becomes

$$\begin{cases} L_{pm}\dfrac{di_e}{dt} + L_m\dfrac{di_c}{dt} + R_{pm}i_e = 0; \\ mg - \dfrac{\partial L_m}{\partial l}i_c i_e = 0; \\ -\dfrac{\partial L_m}{\partial s}i_c i_e = 0; \\ -\dfrac{\partial L_m}{\partial \varphi}i_c i_e = 0. \end{cases} \tag{4.12}$$

As the current i_c is defined by the current generator, using the first equations of set 4.12, the amplitude, I_e, of the current, i_e, can be expressed in terms of the amplitude of current, i_c, in the complex form as follows:

$$I_e = -I\frac{\sqrt{\omega^4 L_{pm}^2 + \omega^2 R_{pm}^2}}{\omega^2 L_{pm}^2 + R_{pm}^2}L_m e^{j\varphi}, \tag{4.13}$$

where $\varphi = \arctan(R_2)/(\omega L_2)$.

Assuming that the condition for the normal operation of the MIS, which requires that the frequency of the feeding current, ω, has to fulfill the inequality $L_{pm}\omega \gg R_{pm}$, is satisfied [33], Equation 4.13 can be reduced to

$$I_e = -I\frac{L_m}{L_{pm}}. \tag{4.14}$$

Taking into account Equation 4.14, set 4.12 can be rewritten as follows:

$$\begin{cases} mg + \dfrac{I^2}{L_{pm}} \dfrac{\partial L_m}{\partial l}\bigg|_{\substack{l=h\\s=0\\\varphi=0}} L_m(h,0,0) = 0; \\[2ex] \dfrac{I^2}{L_{pm}} \dfrac{\partial L_m}{\partial s}\bigg|_{\substack{l=h\\s=0\\\varphi=0}} L_m(h,0,0) = 0; \\[2ex] \dfrac{I^2}{L_{pm}} \dfrac{\partial L_m}{\partial \varphi}\bigg|_{\substack{l=h\\s=0\\\varphi=0}} L_m(h,0,0) = 0. \end{cases} \qquad (4.15)$$

Thus, the obtained set 4.15 describes the static behavior of the MIS. As $L_m(h,0,0)$ and I^2/L_{pm} are positive constants and accounting for Equations 4.9 and 4.10, we can conclude that

$$\dfrac{\partial L_m}{\partial l}\bigg|_{\substack{l=h\\s=0\\\varphi=0}} L_m(h,0,0) = c_l c_0 = -C; \quad \dfrac{\partial L_m}{\partial s}\bigg|_{\substack{l=h\\s=0\\\varphi=0}} = c_s = 0;$$

$$\dfrac{\partial L_m}{\partial \varphi}\bigg|_{\substack{l=h\\s=0\\\varphi=0}} = c_\varphi = 0, \qquad (4.16)$$

where $C = (L_{pm} mg)/(I^2)$.

Hence, the quadratic form of mutual inductance L_m (Equation 4.8) can be modified as

$$L_m(l,s,\varphi) = c_0 - \dfrac{C}{c_0} l + \dfrac{1}{2} c_{ll} l^2 + \dfrac{1}{2} c_{ss} s^2 + \dfrac{1}{2} c_{\varphi\varphi} \varphi^2 \qquad (4.17)$$
$$+ c_{ls} l s + c_{l\varphi} l \varphi + c_{s\varphi} s \varphi.$$

4.3.1.2 MIS Linear Model

Substituting Equation 4.17 into Equation 4.7, the linear model of the MIS, which describes the dynamics of the levitated disk-shaped PM in space, near the equilibrium point can be written as follows [39]:

$$\begin{cases} m\ddot{l} + \mu_l \dot{l} + c_0 \dfrac{I^2}{L_{pm}} \left[\left(c_{ll} + \dfrac{c_l^2}{c_0} \right) l + c_{ls} s + c_{l\varphi} \varphi \right] = F_l; \\[2ex] m\ddot{s} + \mu_s \dot{s} + c_0 \dfrac{I^2}{L_{pm}} \left[c_{ls} l + c_{ss} s + c_{s\varphi} \varphi \right] = F_s; \\[2ex] J\ddot{\varphi} + \mu_\varphi \dot{\varphi} + c_0 \dfrac{I^2}{L_{pm}} \left[c_{l\varphi} l + c_{s\varphi} s + c_{\varphi\varphi} \varphi \right] = M_\varphi. \end{cases} \qquad (4.18)$$

The analysis of MIS designs [19,20,24,34,40] shows that the following geometrical limits can be applied, namely,

$$h \ll r_c, r_{pm}; r_{pm}|\varphi| \ll h \text{ and } |d| \ll r_c, r_{pm}, \quad (4.19)$$

where:
r_c and r_{pm} are the radii of the coil and PM, respectively
$d = r_{pm} - r_c$

Then, the mutual induction as a function of the generalized coordinates can be represented in the integral form as follows:

$$L_m = \frac{\mu_0 r_c}{\pi} \int_0^\pi \left[\ln \frac{8r_c}{\sqrt{l_\lambda^2 + (\rho - r_c)^2}} - 1.92 \right] d\alpha, \quad (4.20)$$

where:
$\rho = \rho(s) \approx \sqrt{(r_{pm}^2 + 2r_{pm} \cdot s \cos\alpha + s^2)}$
$l_\lambda = l_\lambda(l,\varphi) = l - r_{pm} \sin\varphi \cos\alpha$
μ_0 is the magnetic permeability of vacuum

Hence, for the particular MIS design shown in Figure 4.7a, the coefficients 4.9 through 4.11 can be expressed analytically using Equation 4.20 and written as follows:

$$c_{ll} = \left.\frac{\partial^2 L_m}{\partial l^2}\right|_{\substack{l=h \\ s=0 \\ \varphi=0}} = \frac{a}{\pi} \frac{h^2 - d^2}{[h^2 + d^2]^2} \int_0^\pi \alpha \, d\alpha = a \frac{h^2 - d^2}{[h^2 + d^2]^2}; \quad (4.21)$$

$$c_{\varphi\varphi} = \left.\frac{\partial^2 L_m}{\partial \varphi^2}\right|_{\substack{l=h \\ s=0 \\ \varphi=0}} = r_{pm}^2 \frac{a}{\pi} \frac{h^2 - d^2}{[h^2 + d^2]^2} \int_0^\pi \cos^2\alpha \, d\alpha = r_{pm}^2 \frac{a}{2} \frac{h^2 - d^2}{[h^2 + d^2]^2}; \quad (4.22)$$

$$c_{ss} = \left.\frac{\partial^2 L_m}{\partial s^2}\right|_{\substack{l=h \\ s=0 \\ \varphi=0}} = \frac{a}{2} \cdot \frac{d^2(r_{pm} - d) - h^2(r_{pm} + d)}{r_{pm} \cdot [h^2 + d^2]^2}; \quad (4.23)$$

$$c_{s\varphi} = \left.\frac{\partial^2 L_m}{\partial s \partial \varphi}\right|_{\substack{l=h \\ s=0 \\ \varphi=0}} = -a \cdot r_{pm} \frac{d \cdot h}{[h^2 + d^2]^2}; \quad (4.24)$$

$$c_0 = a\left[\ln\frac{8r_c}{\sqrt{h^2 + d^2}} - 1.92\right], c_l = -a\frac{h}{h^2 + d^2}, \quad (4.25)$$

where $a = \mu_0 r_c$.

Since $c_{l\varphi} = c_{ls} = 0$, model 4.18 can be reduced to

$$\begin{cases} m\ddot{l} + \mu_l \dot{l} + c_0 \dfrac{I^2}{L_{pm}}\left(c_{ll} + \dfrac{c_l^2}{c_0}\right)l = F_l; \\ m\ddot{s} + \mu_s \dot{s} + c_0 \dfrac{I^2}{L_{pm}}\left[c_{ss}s + c_{s\varphi}\varphi\right] = F_s; \\ J\ddot{\varphi} + \mu_\varphi \dot{\varphi} + c_0 \dfrac{I^2}{L_{pm}}\left[c_{s\varphi}s + c_{\varphi\varphi}\varphi\right] = M_\varphi. \end{cases} \quad (4.26)$$

4.3.1.3 Stable Levitation

From Equation 4.26, we can write the stiffness matrix of the MIS, which is

$$\mathbf{S} = c_0 \dfrac{I^2}{L_{pm}} \cdot \begin{bmatrix} c_{ll} + \dfrac{c_l^2}{c_0} & 0 & 0 \\ 0 & c_{ss} & c_{s\varphi} \\ 0 & c_{s\varphi} & c_{\varphi\varphi} \end{bmatrix}. \quad (4.27)$$

Applying the Sylvester criterion to Equation 4.27, the condition for the stable levitation of the PM in space can be obtained and written in terms of the coefficients of the quadratic form of the function of mutual inductance:

$$c_{ll} + \dfrac{c_l^2}{c_0} > 0, \left(c_{ll} + \dfrac{c_l^2}{c_0}\right)c_{ss} > 0, c_{ss}c_{\varphi\varphi} > c_{l\varphi}^2. \quad (4.28)$$

Thus, inequalities 4.28 are the general condition for the stable levitation of the disk-shaped PM in space for the design with a one-ring coil under geometrical limits (Equation 4.19).

4.3.2 Analysis of MIS Designs

Although the model of the MIS (Equation 4.26) and the condition for stable levitation (Equation 4.28) are derived for a design consisting of a one-ring-shaped coil and a disk-shaped PM, it can be easily generalized and applied to more complex MIS designs consisting of two or more coaxial ring-shaped coils located on one plane as well as in space [37]. In the framework of linear modeling, the generalization is reached by means of applying the superposition rule. Hence, for a general case, when the MIS design consists of n coaxial coils, model 4.26 can be rewritten as follows:

$$\begin{cases} m\ddot{l} + \mu_l \dot{l} + \dfrac{\sum_{\rho=1}^{n} \pm c_0^\rho I_\rho}{L_{pm}} \left[\sum_{\rho=1}^{n} \pm c_{ll}^\rho I_\rho + \dfrac{\left(\sum_{\rho=1}^{n} \pm c_l^\rho I_\rho\right)^2}{\sum_{\rho=1}^{n} \pm c_0^\rho I_\rho} \right] \cdot l = F_l; \\[2em] m\ddot{s} + \mu_s \dot{s} + \dfrac{\sum_{\rho=1}^{n} \pm c_0^\rho I_\rho}{L_{pm}} \left[\sum_{\rho=1}^{n} \pm c_{ss}^\rho I_\rho \cdot s + \sum_{\rho=1}^{n} \pm c_{s\varphi}^\rho I_\rho \cdot \varphi \right] = F_s; \\[2em] J\ddot{\varphi} + \mu_\varphi \dot{\varphi} + \dfrac{\sum_{\rho=1}^{n} \pm c_0^\rho I_\rho}{L_{pm}} \left[\sum_{\rho=1}^{n} \pm c_{s\varphi}^\rho I_\rho \cdot s + \sum_{\rho=1}^{n} \pm c_{\varphi\varphi}^\rho I_\rho \cdot \varphi \right] = M_\varphi, \end{cases} \quad (4.29)$$

where:

ρ denotes the coil number as the subscript in the current I_ρ and the superscript in coefficients of mutual induction $\left(c_0^\rho, c_1^\rho, c_{nw}^\rho \left(n = l, s, \varphi; w = l, s, \varphi\right)\right)$

I_ρ is the amplitude of the current within coil with ρ number, the ± sign defines the direction of the I_ρ current

In assuming that phase shifts between the currents can be 0° and 180° only, let us assign the + sign to the current flowing in the clockwise direction, while the − sign to the counterclockwise direction.

Then, as in Section 4.3.1.3, the condition for stable levitation can be written in detail as follows:

$$\begin{cases} \dfrac{\sum_{\rho=1}^{n} \pm c_0^\rho I_\rho}{L_{pm}} > 0; \quad \sum_{\rho=1}^{n} \pm c_{ll}^\rho I_\rho + \dfrac{\left(\sum_{\rho=1}^{n} \pm c_l^\rho I_\rho\right)}{\sum_{\rho=1}^{n} \pm c_0^\rho I_\rho} > 0, \\[2em] \sum_{\rho=1}^{n} \pm c_{ss}^\rho I_\rho > 0, \\[1.5em] \sum_{\rho=1}^{n} \pm c_{\varphi\varphi}^\rho I_\rho > 0, \\[1.5em] \sum_{\rho=1}^{n} \pm c_{ss}^\rho I_\rho \cdot \sum_{\rho=1}^{n} \pm c_{\varphi\varphi}^\rho I_\rho > \left(\sum_{\rho=1}^{n} \pm c_{s\varphi}^\rho I_\rho\right)^2. \end{cases} \quad (4.30)$$

Micromachined Inductive Contactless Suspension

Now, using model 4.29 and the condition for stable levitation (Equation 4.30), a procedure for analyzing MIS designs can be suggested as follows. In the first step, condition 4.30 is applied to a proposed MIS design. If, in the proposed design, stable levitation is possible, then the stability domain as a function of the geometrical parameters of the MIS is defined and mapped. In the second step, within the defined stability domain the components of the MIS stiffness are calculated and mapped.

4.3.2.1 Preliminary Analysis of Stability of MIS Based on Plane Coils

Let us apply the previously mentioned models for the simplest MIS designs, which are shown in Figures 4.1. and 4 7b, in order to study their stability. A one-plane coil design of the MIS shown in Figure 4.7b is unstable and as a result it is not of practical use and so the analysis of its stability is left here as an exercise for the reader. (Also the stability analysis for a one-plane coil MIS design can be found in [37].)

At the beginning of the stability study of the two-plane coils MIS design, let us define its geometrical parameters as shown in Figure 4.9. It is assumed that the currents in the coils have the same amplitudes but opposite directions of flowing. Then, accounting for Equations 4.21 through 4.25 the coefficients of mutual induction become as follows:

$$c_{ll} + \frac{c_l^2}{c_0} = a_s \frac{h^2-(d-c)^2}{\left[h^2+(d-c)^2\right]^2} - a_l \frac{h^2-d^2}{\left[h^2+d^2\right]^2} + \varepsilon h^2 \left(\frac{a_s}{h^2+(d-c)^2} - \frac{a_l}{h^2+d^2}\right)^2, \quad (4.31)$$

$$c_{ss} = \frac{a_s}{2} \cdot \frac{(d-c)^2(r_{pm}+c-d)-h^2(r_{pm}-c+d)}{r_{pm} \cdot [h^2+(d-c)^2]^2} - \frac{a_l}{2} \cdot \frac{d^2(r_{pm}-d)-h^2(r_{pm}+d)}{r_{pm} \cdot [h^2+d^2]^2}, \quad (4.32)$$

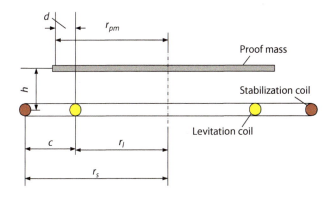

FIGURE 4.9
Two coils MIS design: r_{pm}, r_l, and r_s are radii of the PM, levitation coil, and stabilization coil, respectively; $d = r_{pm} - r_l$, $c = r_s - r_l$.

$$C_{\varphi\varphi} = r_{pm}^2 \left[\frac{a_s}{2} \frac{h^2 - (d-c)^2}{\left[h^2 + (d-c)^2\right]^2} - \frac{a_l}{2} \frac{h^2 - d^2}{\left[h^2 + d^2\right]^2} \right], \tag{4.33}$$

$$C_{s\varphi} = r_{pm}h \left[a_s \frac{(c-d)}{\left[h^2 + (d-c)^2\right]^2} + a_l \frac{d}{\left[h^2 + d^2\right]^2} \right], \tag{4.34}$$

where:
$a_s = \mu_0 r_s$
$a_l = \mu_0 r_l$

$$\varepsilon = \left\{ a_s \left[\ln \frac{8r_s}{\sqrt{h^2 + (d-c)^2}} - 1.92 \right] + a_l \left[\ln \frac{8r_l}{\sqrt{h^2 + d^2}} - 1.92 \right] \right\}^{-1}. \tag{4.35}$$

Analysis of Equations 4.31 through 4.42 shows that d ($d = r_{pm} - r_l$) and h can be considered as independent variables for the stability study. Applying condition 4.28, the stability domain can be mapped as shown in Figure 4.10. Thus, in the two-plane coils MIS design, stable levitation is possible for the size of the PM defined in the range from $c\beta/(1+\beta)$ to $c\beta/(\beta + \sqrt{\beta})$, where $\beta = r_l/r_s$.

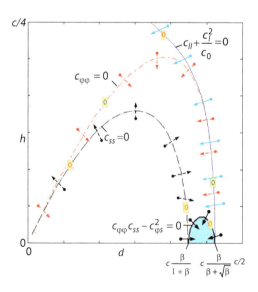

FIGURE 4.10
The stability domain of d and h shown in blue-gray: arrows show the change of sign from minus to plus of the respective equality, when its boundary is crossed in the indicated direction; $\beta = r_l/r_s$.

4.3.2.2 Analysis of MIS Design Based on Two 3-D Coils

Now, let us study the MIS based on 3-D coils design (3-D MIS) presented in Figure 4.4. Although, this design consists of two coils, which is similar to the two-plane coils design previously considered; however, 3-D coils allow us to add windings for each coil. The winding is characterized by the pitch and a number of turns. Figure 4.11 presents the scheme of 3-D MIS, which reflects the particularities of the winding, to define the geometrical parameters of the design under consideration.

Since there are two ac feeding the coils with amplitudes I_S and I_L in the stabilization and levitation coil, respectively, model 4.29 can be rewritten as follows:

$$\begin{cases} m\ddot{l} + \mu_l \dot{l} + \dfrac{c_0^S I_S \pm c_0^L I_L}{L_{PM}} \left(c_{ll}^S I_S \pm c_{ll}^L I_L + \dfrac{\left(c_l^S I_S \pm c_l^L I_L\right)^2}{c_0^S I_S \pm c_0^L I_L} \right) \cdot l = F_l; \\ m\ddot{s} + \mu_s \dot{s} + \dfrac{c_0^S I_S \pm c_0^L I_L}{L_{PM}} \left[\left(c_{ss}^S I_S \pm c_{ss}^L I_L\right) \cdot s + \left(c_{s\varphi}^S I_S \pm c_{s\varphi}^L I_L\right) \cdot \varphi \right] = F_s; \\ J\ddot{\varphi} + \mu_\varphi \dot{\varphi} + \dfrac{c_0^S I_S \pm c_0^L I_L}{L_{PM}} \left[\left(c_{s\varphi}^S I_S \pm c_{s\varphi}^L I_L\right) \cdot s + \left(c_{\varphi\varphi}^S I_S \pm c_{\varphi\varphi}^L I_L\right) \cdot \varphi \right] = M_\varphi, \end{cases} \quad (4.36)$$

where the coefficient superscripts L and S correspond to the stabilization and levitation coil, respectively. Using Equations 4.21 through 4.25 and the geometry of the design presented in Figure 4.11, the coefficients can be calculated by

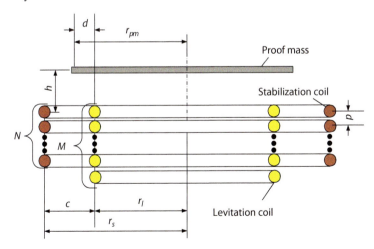

FIGURE 4.11
3-D coils MIS design: p is the coils pitch of winding; N and M are numbers of winding turns for the stabilization and levitation coil, respectively.

$$c_0^S = \sum_{v=0}^{N-1} \frac{a_s}{\gamma_s}\left[\ln\frac{8r_s}{\sqrt{(h+v\cdot p)^2+(d-c)^2}}-1.92\right];$$

$$c_0^L = \sum_{v=0}^{M-1} \frac{a_l}{\gamma_l}\left[\ln\frac{8r_l}{\sqrt{(h+v\cdot p)^2+d^2}}-1.92\right];$$
(4.37)

$$c_l^S = \sum_{v=0}^{N-1} -\frac{a_s}{\gamma_s}\frac{(h+v\cdot p)}{(h+v\cdot p)^2+(d-c)^2};\ c_l^L = \sum_{v=0}^{M-1} -\frac{a_l}{\gamma_l}\frac{(h+v\cdot p)}{(h+v\cdot p)^2+d^2};$$ (4.38)

$$c_{ll}^S = \sum_{v=0}^{N-1} \gamma_s a_s \frac{(h+v\cdot p)^2-(d-c)^2}{\left[(h+v\cdot p)^2+(d-c)^2\right]^2};\ c_{ll}^L = \sum_{v=0}^{M-1} \gamma_l a_l \frac{(h+v\cdot p)^2-d^2}{\left[(h+v\cdot p)^2+d^2\right]^2};$$ (4.39)

$$c_{ss}^S = \sum_{v=0}^{N-1} \gamma_s \frac{a_s}{2}\cdot\frac{(d-c)^2 r_s-(h+v\cdot p)^2(r_s-2(c-d))}{(r_s-(c-d))\cdot[(h+v\cdot p)^2+(d-c)^2]^2};$$

$$c_{ss}^L = \sum_{v=0}^{M-1} \gamma_l \frac{a_l}{2}\cdot\frac{d^2 r_l-(h+v\cdot p)^2(r_l+2d)}{(r_l+d)\cdot[(h+v\cdot p)^2+d^2]^2};$$
(4.40)

$$c_{\varphi\varphi}^S = \sum_{v=0}^{N-1} \gamma_s \frac{a_s\cdot(r_l+d)^2}{2}\cdot\frac{(h+v\cdot p)^2-(d-c)^2}{\left[(h+v\cdot p)^2+(d-c)^2\right]^2};$$

$$c_{\varphi\varphi}^L = \sum_{v=0}^{M-1} \gamma_l \frac{a_l\cdot(r_l+d)^2}{2}\cdot\frac{(h+v\cdot p)^2-d^2}{\left[(h+v\cdot p)^2+d^2\right]^2};$$
(4.41)

$$c_{s\varphi}^S = \sum_{v=0}^{N-1} a_s\gamma_{sl}\gamma_s \frac{(r_l+d)(h+v\cdot p)(c-d)}{\left[(h+v\cdot p)^2+(d-c)^2\right]^2};\ c_{s\varphi}^L = \sum_{v=0}^{M-1} -a_l\gamma_{sl}\gamma_l \frac{(r_l+d)(h+v\cdot p)d}{\left[(h+v\cdot p)^2+d^2\right]^2};$$ (4.42)

where:
- μ_0 = $4\pi \times 10^{-7}$ H/m is the magnetic permeability of vacuum
- a_s = $\mu_0 r_s$
- a_l = $\mu_0 r_l$, r_s
- r_l are the radii of the stabilization and levitation coil, respectively
- p is the coils pitch of winding
- c = $r_s - r_l$, r_{pm} is the radius of the PM
- d = $r_{pm} - r_l$
- N and M are numbers of winding turns for the stabilization and levitation coil, respectively
- γ_{sl}, γ_s, and γ_l are coefficients of similarity

Coefficients of similarity are introduced to rearrange the real contribution of each coefficient of the quadratic form and adapt model 4.36 to the real condition of the suspension operation. This becomes necessary due in reality to the induced eddy current that is distributed along the PM surface, a distribution that is frequency dependent. However, analytical model 4.36 is obtained by considering the suspension as a lumped system.

From Equation 4.30, the condition for stable levitation is

$$c_0^S I_S \pm c_0^L I_L > 0; \quad c_{ll}^S I_S \pm c_{ll}^L I_L + \frac{\left(c_l^S I_S \pm c_l^L I_L\right)^2}{c_0^S I_S \pm c_0^L I_L} > 0; \quad (4.43a)$$

$$c_{ss}^S I_S \pm c_{ss}^L I_L > 0; \quad (4.43b)$$

$$c_{\varphi\varphi}^S I_S \pm c_{\varphi\varphi}^L I_L > 0; \quad (4.43c)$$

$$\left(c_{\varphi\varphi}^S I_S \pm c_{\varphi\varphi}^L I_L\right) \cdot \left(c_{ss}^S I_S \pm c_{ss}^L I_L\right) > \left(c_{s\varphi}^S I_S \pm c_{s\varphi}^L I_L\right)^2. \quad (4.43d)$$

Let us consider the 3-D MIS design with the geometrical parameters shown in Table 4.2. An experimental study of the 3-D MIS prototype with the same parameters was reported in [67].

Using the foregoing set of inequalities (Section 4.3.2.2), the stability of the 3-D MIS design is studied. Analysis of Equations 4.37 through 4.42 shows that d ($d = r_{pm} - r_l$) and h can be considered as independent variables for this study. Note that the currents in the coils have a 180° phase shift between them, which corresponds to the minus sign in Equation 4.36. Also, we recommend the following values for coefficients of similarity, $\gamma_s = 0.3$, $\gamma_l = 1.2$, and $\gamma_{sl} = 0.14$, as was previously explained.

Substituting the parameters of the prototype shown in Table 4.2 and the values for the currents in the coils, $I_S = 0.106$ A and $I_L = 0.11$ A taken from [38], into Equations 4.37 through 4.42, we find the boundaries of inequalities 4.43a through 4.43d and the stability domain of d and h. This stability domain is mapped in Figure 4.12. The analysis of Figure 4.12 shows

TABLE 4.2

Parameters of the 3-D MIS Design

Radius of the levitation coil, r_l	1000 µm
Radius of the stabilization coil, r_s	1900 µm
The coils pitch of winding, p	25 µm
Number of windings for stabilization coil, N	12
Number of windings for levitation coil, M	20
Radius of PM, r_{pm}	1600 µm
Thickness of PM	25 µm

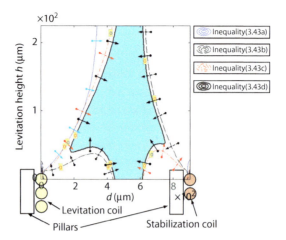

FIGURE 4.12
Boundaries of inequalities 4.43a through 4.43d and the stability domain of d and h filled by the blue-gray color: the arrows show the change of sign from minus to plus of the respective inequality, when its bound is crossed in the shown direction.

that stable levitation in such a prototype is possible. However, solving the inequalities 4.43a through 4.43d for the case when there is no phase shift between the currents in the two coils, one can demonstrate that there is no stability domain, that is, stable levitation is impossible. This fact agrees with the experimental observation, which confirmed that stable levitation in the prototype under consideration is only possible when the currents have a 180° phase shift.

Now, let us calculate the stiffnesses of the 3-D MIS prototype, namely, k_l, k_s, and k_φ, which can be defined according to model 4.18 as follows:

$$k_l = \frac{c_0^S I_S - c_0^L I_L}{L_{PM}} \left(c_{ll}^S I_S - C_{ll}^L I_L + \frac{\left(c_l^S I_S - c_l^L I_L\right)^2}{c_0^S I_S - c_0^L I_L} \right); \quad (4.44)$$

$$k_s = \frac{c_0^S I_S - c_0^L I_L}{L_{PM}} \left(c_{ss}^S I_S - c_{ss}^L I_L \right); \quad (4.45)$$

$$k_\varphi = \frac{c_0^S I_S - c_0^L I_L}{L_{PM}} \left(c_{\varphi\varphi}^S I_S - c_{\varphi\varphi}^L I_L \right), \quad (4.46)$$

where the self-inductance of the PM, L_{PM}, can be calculated as [41]

$$L_{PM} = \mu_0 (r_l + d) \left[\ln \frac{8(r_l + d)}{\delta} - 1.92 \right], \quad (4.47)$$

Micromachined Inductive Contactless Suspension 127

where δ is the effective width in which the maximum of an induced eddy current within the PM is distributed.

As shown in [42], the effective width can be evaluated as $\delta = 0.05\ldots0.1 \cdot r_{pm}$. Using Equations 4.47 through 4.49, the distribution of stiffnesses within the stability domain of a 3-D MIS prototype is mapped and presented in Figure 4.13. Figure 4.13 also shows the positioning of the coils and the PM relative to the stability domain. The region of interest for the calculation is located at the outer edge of the PM. In this particular case, the levitation height is 114 μm and the diameter of the levitated PM is 3.2 mm. Hence, the coordinates for calculation are $d = 600$ μm and $h = 114$ μm. The results

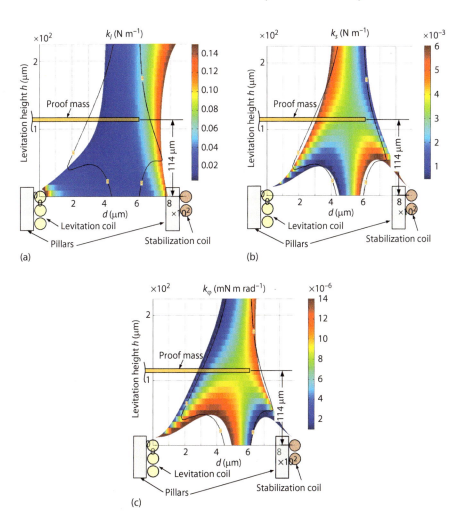

FIGURE 4.13
Distribution of the stiffness components, namely, (a) k_l, (b) k_s, and (c) k_φ, within the stability domain.

TABLE 4.3
Comparison of Suspension Stiffness from Modeling and Experiment Results

Stiffnesses	Measured Values	Calculated Values by Model 4.36
k_s (N·m^{-1})	3.0×10^{-3}	3.0×10^{-3}
k_l (N·m^{-1})	4.5×10^{-2}	4.2×10^{-2}
k_φ (N·m·rad^{-1})	1.5×10^{-8}	0.8×10^{-8}

of both measurements and calculation are shown in Table 4.3. Note that the outer edge of the PM is located within the stability domain. This fact confirms theoretically that the levitation of the PM under consideration is stable. The analysis of Table 4.3 shows that model 4.18 agrees well with the measurements.

Acknowledgments

Dr. Zhiqiu Lu gratefully acknowledges the support from the Siemens-DAAD Scholarship. Dr. Kirill Poletkin acknowledges with thanks the support of the Alexander von Humboldt Foundation. Dr. Vlad Badilita kindly acknowledges support from the German Research Foundation (DFG) through project number BA 4275/2-1.

References

1. L-S Fan, Y-C Tai, and RS Muller, IC-processed electrostatic micromotors, *Sensors and Actuators*, vol. 20, no. 1, pp. 41–47, 1989.
2. U Wallrabe, P Bley, B Krevet, W Menz, and J Mohr, Theoretical and experimental results of an electrostatic micro motor with large gear ratio fabricated by the liga process, in *Micro Electro Mechanical Systems, 1992, MEMS'92, Proceedings. An Investigation of Micro Structures, Sensors, Actuators, Machines and Robot. IEEE*. IEEE, 1992, Travemunde, Germany, pp. 139–140.
3. B Boxenhorn and P Greiff, Monolithic silicon accelerometer, *Sensors and Actuators A: Physical*, vol. 21, no. 1, pp. 273–277, 1990.
4. P Greiff, B Boxenhorn, T King, and L Niles, Silicon monolithic micromechanical gyroscope, in *Solid-State Sensors and Actuators, 1991. Digest of Technical Papers, TRANSDUCERS'91., 1991 International Conference on*. IEEE, 1991, San Francisco, CA. pp. 966–968.

5. C Shearwood and RB Yates, Development of an electromagnetic microgenerator, *Electronics Letters*, vol. 33, no. 22, pp. 1883–1884, 1997.
6. H Fujita, Microactuators and micromachines, *Proceedings of the IEEE*, vol. 86, no. 8, pp. 1721–1732, 1998.
7. RP Feynman, There's plenty of room at the bottom, *Engineering and Science*, vol. 23, no. 5, pp. 22–36, 1960.
8. EH Brandt, Levitation in physics, *Science*, vol. 243, no. 4889, pp. 349–355, 1989.
9. J Jin, T Higuchi, and M Kanemoto, Electrostatic levitator for hard disk media, *Industrial Electronics, IEEE Transactions on*, vol. 42, no. 5, pp. 467–473, 1995.
10. JU Jeon and T Higuchi, Electrostatic suspension of dielectrics, *Industrial Electronics, IEEE Transactions on*, vol. 45, no. 6, pp. 938–946, 1998.
11. R Toda, N Takeda, T Murakoshi, S Nakamura, and M Esashi, Electrostatically levitated spherical 3-axis accelerometer, in *Micro Electro Mechanical Systems, 2002. The Fifteenth IEEE International Conference on*. IEEE, 2002, Las Vegas, NV. pp. 710–713.
12. F Han, B Sun, L Li, and Q Wu, Performance of a sensitive micromachined accelerometer with an electrostatically suspended proof mass, *IEEE Sensors Journal*, vol. 15, pp. 209–217, 2015.
13. B Damrongsak and M Kraft, A micromachined electrostatically suspended gyroscope with digital force feedback, in *Sensors, 2005 IEEE*. IEEE, 2005, Irvine, CA. pp. 401–404.
14. B Damrongsak, M Kraft, S Rajgopal, and M Mehregany, Design and fabrication of a micromachined electrostatically suspended gyroscope, *Proceedings of the Institution of Mechanical Engineers, Part C: Journal of Mechanical Engineering Science*, vol. 222, no. 1, pp. 53–63, 2008.
15. Y Su, Z Xiao, Z Ye, and K Takahata, Micromachined graphite rotor based on diamagnetic levitation, *Electron Device Letters, IEEE*, vol. 36, no. 4, pp. 393–395, 2015.
16. T Murakoshi, Y Endo, K Fukatsu, S Nakamura, and M Esashi, Electrostatically levitated ring-shaped rotational gyro/accelerometer, *Japanese Journal of Applied Physics*, vol. 42, no. 4B, pp. 2468–2472, 2003.
17. S Nakamura, MEMS inertial sensor toward higher accuracy & multi-axis sensing, in *Sensors, 2005 IEEE*. IEEE, 2005, Irvine, CA, pp. 939–942.
18. C Ruffert, R Gehrking, B Ponick, and HH Gatzen, Magnetic levitation assisted guide for a linear micro-actuator, *Magnetics, IEEE Transactions on*, vol. 42, no. 11, pp. 3785–3787, 2006.
19. C Shearwood, KY Ho, CB Williams, and H Gong, Development of a levitated micromotor for application as a gyroscope, *Sensors and Actuators A-Physical*, vol. 83, no. 1–3, pp. 85–92, 2000.
20. W Zhang, W Chen, X Zhao, X Wu, W Liu, X Huang, and S Shao, The study of an electromagnetic levitating micromotor for application in a rotating gyroscope, *Sensors and Actuators A-Physical*, vol. 132, no. 2, pp. 651–657, 2006.
21. N-C Tsai, W-M Huan, and C-W Chiang, Magnetic actuator design for single-axis micro-gyroscopes, *Microsystem Technologies*, vol. 15, pp. 493–503, 2009.
22. I Sari and M Kraft, A micro electrostatic linear accelerator based on electromagnetic levitation, in *Solid-State Sensors, Actuators and Microsystems Conference (TRANSDUCERS), 2011 16th International*. IEEE, 2011, Beijing, China, pp. 1729–1732.
23. I Sari and M Kraft, A MEMS linear accelerator for levitated micro-objects, *Sensors and Actuators A-Physical*, vol. 222, pp. 15–23, 2015.

24. C Shearwood, CB Williams, PH Mellor, RB Yates, MRJ Gibbs, and AD Mattingley, Levitation of a micromachined rotor for application in a rotating gyroscope, *Electronics Letters*, vol. 31, no. 21, pp. 1845–1846, 1995.
25. V Badilita, S Rzesnik, K Kratt, and U Wallrabe, Characterization of the 2nd generation magnetic microbearing with integrated stabilization for frictionless devices, in *Solid-State Sensors, Actuators and Microsystems Conference (TRANSDUCERS), 2011 16th International. 2011*, Beijing, China, pp. 1456–1459, IEEE.
26. D Garmire, H Choo, R Kant, S Govindjee, CH Sequin, RS Muller, and J Demmel, Diamagnetically levitated mems accelerometers, in *Solid-State Sensors, Actuators and Microsystems Conference, 2007. TRANSDUCERS 2007. International*. IEEE, 2007, Lyon, France, pp. 1203–1206.
27. W Liu, W-Y Chen, W-P Zhang, X-G Huang, and Z-R Zhang, Variablecapacitance micromotor with levitated diamagnetic rotor, *Electronics Letters*, vol. 44, no. 11, pp. 681–683, 2008.
28. Y-K Kim, M Katsurai, and H Fujita, A superconducting actuator using the meissner effect, *Sensors and Actuators*, vol. 20, no. 1, pp. 33–40, 1989.
29. TA Coombs, I Samad, D Ruiz-Alonso, and K Tadinada, Superconducting micro-bearings, *Applied Superconductivity, IEEE Transactions on*, vol. 15, no. 2, pp. 2312–2315, 2005.
30. K Liu, W Zhang, W Liu, W Chen, K Li, F Cui, and S Li, An innovative micro-diamagnetic levitation system with coils applied in micro-gyroscope, *Microsystem Technologies*, vol. 16, no. 3, pp. 431–439, 2010.
31. K Poletkin, Z Lu, U Wallrabe, and V Badilita, JMEMS Letters A new hybrid micromachined contactless suspension with linear and angular positioning and adjustable dynamics, *Journal of Microelectromechanical Systems*, vol. 24, no. 5, pp. 1248–1250, 2015.
32. O Cugat, J Delamare, and G Reyne, Magnetic microactuators and systems (magmas), *Magnetics, IEEE Transactions on*, vol. 39, no. 6, pp. 3607–3612, 2003.
33. KV Poletkin, AI Chernomorsky, and C Shearwood, A proposal for micromachined accelerometer, base on a contactless suspension with zero spring constant, *IEEE Sensors Journal*, vol. 12, no. 07, pp. 2407–2413, 2012.
34. CB Williams, C Shearwood, and PH Mellor, Modeling and testing of a frictionless levitated micromotor, *Sensors and Actuators A-Physical*, vol. 61, pp. 469–473, 1997.
35. K Kratt, V Badilita, T Burger, JG Korvink, and U Wallrabe, A fully MEMS-compatible process for 3D high aspect ratio micro coils obtained with an automatic wire bonder, *Journal of Micromechanics and Microengineering*, vol. 20, pp. 015021, 2010.
36. Z Lu, K Poletkin, U Wallrabe, and V Badilita, Performance characterization of micromachined inductive suspensions based on 3D wirebonded microcoils, *Micromachines*, vol. 5, no. 4, pp. 1469–1484, 2014.
37. K Poletkin, A Chernomorsky, C Shearwood, and U Wallrabe, A qualitative analysis of designs of micromachined electromagnetic inductive contactless suspension, *International Journal of Mechanical Sciences*, vol. 82, pp. 110–121, 2014.
38. Z Lu, K Poletkin, B den Hartogh, U Wallrabe, and V Badilita, 3D micromachined inductive contactless suspension: Testing and modeling, *Sensors and Actuators A Physical*, vol. 220, pp. 134–143, 2014.

39. K Poletkin, AI Chernomorsky, C Shearwood, and U Wallrabe, An analytical model of micromachined electromagnetic inductive contactless suspension, in *The ASME 2013 International Mechanical Engineering Congress & Exposition*, San Diego, CA, 2013, pp. V010T11A072–V010T11A072, ASME.
40. V Badilita, M Pauls, K Kratt, and U Wallrabe, Contactless magnetic micro-bearing based on 3D solenoidal micro-coils for advanced powerMEMS components, in *Proceedings of PowerMEMS 2009*, 2009, pp. 87–90.
41. FW Grover, *Inductance Calculations: Working Formulas and Tables*. Chicago, IL: Dover, 2004.
42. Z Lu, F Jia, JG Korvink, U Wallrabe, and V Badilita, Design optimization of an electromagnetic microlevitation system based on copper wirebonded coils, in *2012 Power MEMS Proceedings of PowerMEMS*, Atlanta, GA, 2012, pp. 363–366.
43. KV, Poletkin A novel hybrid contactless suspension with adjustable spring constant, in *Solid-State Sensors, Actuators and Microsystems Conference (TRANSDUCERS), 2017 19th International. IEEE, 2017*, Taiwan. (In press).
44. KV, Poletkin, Z, Lu, U Wallrabe, J Korvink, and V Badilita, A qualitative technique to study stability and dynamics of micro-machined inductive contactless suspensions, in *Solid-State Sensors, Actuators and Microsystems Conference (TRANSDUCERS), 2017 19th International . IEEE, 2017*, Taiwan. (in press).
45. KV Poletkin, Z Lu, A Moazenzadeh, SG Mariappan, J Korvink, U Wallrabe, and V Badilita, Polymer magnetic composite core boosts performance of 3-D micromachined inductive contactless suspension. *IEEE Magnetics Letters*. vol. 7, pp. 1–3 2016.

5

Application of Magnetic Sensors for Ecological Monitoring of Stationary Ferromagnetic Masses from On Board Mobile Platforms

Alexander I. Chernomorsky, Vyacheslav E. Plekhanov, and Vladimir N. Maximov

CONTENTS

5.1 Introduction 134
5.2 Monitoring Local Stationary Ferromagnetic Masses from On Board the AUV 135
    5.2.1 Particularities of Monitoring Local Stationary Ferromagnetic Masses from On Board Mobile Platforms 135
    5.2.2 Estimation of Magnetic Conditions On Board the AUV 135
    5.2.3 Analysis of FM Magnetic Field and Estimation of Distance to Area of FM Location 137
    5.2.4 Overview of Typical MS Magnetometers 139
5.3 Calibration of Vector Magnetic Sensors: Magnetometers 140
    5.3.1 First Method for VMS Calibration 142
    5.3.2 Second Method for VMS Calibration 144
        5.3.2.1 Algorithm for Parameters Estimation of Background Magnetic Field Model 144
        5.3.2.2 Optimization Algorithm for Changing AUV Angular Orientation on Trajectory of Its Motion 145
5.4 Determining Location of FM Relative to Mobile Platform and Its Transportation to Area of FMs Location 151
    5.4.1 Determining Location of FM 151
    5.4.2 Algorithms for Determining FM Location Relative to AUV ... 153
    5.4.3 Transportation of MP to the Area of FM Location 156
5.5 Electromagnetic System for Positioning and Azimuth Angular Orientation of Ground Mobile Platform 156
    5.5.1 EONS Configuration 157
    5.5.2 Orientation and Positioning by Means of EONS 160
References 162

ABSTRACT In this chapter, the application of magnetic sensors (MS), in particular vector magnetic sensors (VMSs), for solving problems related to the ecological monitoring of local stationary ferromagnetic masses (FMs; e.g., sunken watercrafts and underwater structures) on board autonomous robotic mobile platforms (MPs) is discussed. An autonomous underwater vehicle—robots (AUV) [1,2] and a ground uniaxial wheel transport platform (UWTP) [3,4] are considered as such MPs. As a rule, the monitoring of FMs aboard these MPs is reduced to the following procedure. First, the coordinates of an FM relative to the MP are determined. Second, the MP is transported to the location of the FM. Obviously, in order to perform this monitoring, inertial information on the MP is essential for its navigation and orientation.

5.1 Introduction

In Section 5.2, the local stationary monitoring of *ferromagnetic masses* (FMs), in particular by means of a *vector magnetic sensor* (VMS) mounted on the *autonomous underwater vehicle—robots* (AUV), is studied. An estimation of the magnetic conditions on board the AUV is given. Also in this section, the characteristic of a magnetic field induced by FMs is analyzed. Based on the results of the analysis, the distance to the area of the FM's location is estimated. In addition, a review of existing *magnetic sensors* (MS) magnetometers is presented.

In Section 5.3, the calibration methods of the VMS on board a stationary and moving AUV are described. The proposed methods provide the possibility for compensation of magnetic disturbances due to magnetic field sources on board the mobile platform (MP). In particular, an algorithm that optimizes the profile of changing the AUV angular orientation on a trajectory of its motion is proposed. The proposed algorithm is used to maximize the accuracy of determination of the parameters of the model describing the induction vector of the background magnetic field.

In Section 5.4, a determination of FM coordinates relative to an MP (AUV) by means of VMSs mounted onboard the MP is studied. It is important to note that the well-known method for solving this problem based on employing a magnetic field sensor and a magnetic gradiometer has the difficulty of implementing its hardware on board the MP. Because of this, it is proposed to use only one VMS to perform the measurements of the gradient of a magnetic field and its induction with further determining coordinates of an FM. Moreover, the gradient of a magnetic field induced by an FM is calculated by measuring the increment of the vector of magnetic induction upon moving the MP along a known trajectory, the parameters of which are defined by the MP navigation system. In order to transport the MP to the area where the

FM is located, guidance of the MP is realized by utilizing the information received on the MP trajectory about the gradient vector of the module of the FM magnetic field.

In Section 5.54, the operational principle and the results of the development of algorithms for measuring the electromagnetic orientation and navigation system (EONS) of the MP, such as a UWTP, which is used for monitoring FMs located on local land surfaces, are discussed. As an MS, UWTP uses the inductive MS.

5.2 Monitoring Local Stationary Ferromagnetic Masses from On Board the AUV

In this section, the magnetic conditions on board the AUV are considered. The characteristics of a magnetic field induced by an FM are analyzed. Also, an estimation of the distance to the area of the FM location is given. In addition, a review of MS, which are suitable for monitoring FMs, is provided.

5.2.1 Particularities of Monitoring Local Stationary Ferromagnetic Masses from On Board Mobile Platforms

As a rule, the application of MS for monitoring a local FM is accompanied by a high level of magnetic disturbance compared with the small values of a magnetic field induced by an FM located an acceptable distance for its detection. Magnetic disturbance is due to the background magnetic field of the MP, which is the sum of the permanent magnetic field of the MP itself and the magnetic field of inductively magnetized ferromagnetic elements of the MP design, as well as the Earth's magnetic field (EMF).

Thus, upon monitoring local FMs, initially, the background magnetic field should be restored (determined) on board the MP. Then, the induction of a magnetic field generated by FMs on a trajectory of the MP motion can be defined as the difference between the current measurement vector of magnetic field induction and the restored vector of the background magnetic field. To restore a vector of the background magnetic field on board the MP, information about the current linear coordinates and angular orientation of the VMS mounted on the MP is necessary.

5.2.2 Estimation of Magnetic Conditions On Board the AUV

The background magnetic field in the monitoring area can be represented as the sum of its stationary and transient components. The stationary component can be identified during the calibration of a VMS, while the transient component defines the overall error of monitoring. In its turn, the stationary component of the EMF is the sum of its normal and anomalous fields. The normal field is defined by the global structure of the EMF and is changed within a range of 30–60 µT. The intensity of the anomalous field of the EMF and its distribution in space are dependent on its magnetic properties, geological structure, and the depth of occurrence of non-magnetic rocks of the Earth's crust. The value of the anomalous part of the EMF can be up to 1 µT.

As previously mentioned, the other source of a stationary magnetic field is the AUV itself, namely, the permanent and inductively magnetized ferromagnetic elements of its design, as well as the electrical loads. Permanent magnetization of elements of the AUV design occurs during the fabrication of these elements and their exploitation. It depends essentially on the various mechanical and thermal treatments of these elements. In general, inductive magnetization of ferromagnetic elements of the AUV design is dependent on their magnetic permeability, geometrical shape, and orientation in the external magnetic field. The magnetic permeability, in turn, as well as the residual magnetization, is dependent on the chemical composition and processing technology of the design elements. Usually, structural steels have small values of magnetic permeability. This explains the fact that a magnetic field of inductive magnetization is much weaker than one of permanent magnetization of elements of the AUV design. Also note that magnetic fields generated by inductively magnetized elements of an MP design are hardly changed with mechanical and thermal influences. Typical levels of stationary magnetic fields generated by various sources, which have been previously discussed, are depicted in Figure 5.1

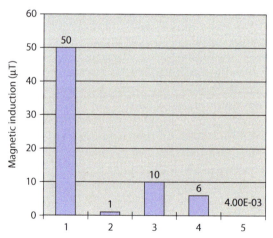

1: normal EMF; 2: EMF anomaly; 3: permanent platform magnetic field; 4: platform inductive magnetic field; 5: magnetometer noise

FIGURE 5.1
Typical levels of stationary magnetic fields on board an AUV.

Transient magnetic fields on board the AUV can be viewed mainly as the sum of the transient magnetic fields of the EMF and the AUV. A transient magnetic field of the EMF includes secular EMF variations (0.02 µT), solar and lunar daily EMF variations (0.02 µT), and variations generated by magnetic storms (0.1 µT). The AUV's own magnetic fields make the main contribution to the transient fields. Indeed, the transient magnetic field of the AUV is determined by the transient magnetic fields of inductively magnetized ferromagnetic elements of the AUV, the transient magnetic fields of electrical loads, as well as the moving magnetic elements of the AUV.

Transient inductive magnetization is mostly defined by the transient magnetic field of the EMF. In a general case, transient magnetic fields are defined by electrical loads. In turn, transient processes in electrical loads are dependent on the area of electric circuits occupied by electrical loads and the location of these electric circuits relative to the MS. The largest value of magnetic disturbance created by the electric circuits is due to single-wire electric systems, where the housing of the AUV is used as an electric conductor. In this case, the area of the electric circuits becomes significant. The induction of the single-wire electric system is decreased inversely proportional to the first power of the distance from an electric source. Also, a significant disturbance can be created by microchips located near the MS.

5.2.3 Analysis of FM Magnetic Field and Estimation of Distance to Area of FM Location

It is assumed that an FM has the residual magnetization, B_{res}, and a significant value of the relative magnetic permeability µ (order of 10 and larger). Some typical parameters of monitoring FMs (sunken watercrafts, underwater structures, etc.) such as their dimensions, B_{res} and µ, as well as relationships that allow to estimate the value of an FM magnetic field depending on the distance to the FM are shown in Table 5.1. The dependences of magnetic fields induced by FMs with residual and inductive magnetizations on the distance to these FMs are shown in Figures 5.2 and 5.3 for B_{res} = 0.5 and 10 µT, respectively. Also, these figures depict the average values of the transient background magnetic field (0.2 µT) and the noise of the VMS ($4*10^{-4}$ µT). The distance to the area of FM location is estimated by points of crossing the dependence of magnetic field induction with straight lines corresponding to the values of a transient background magnetic field and the noise of the VMS, as shown in Figures 5.2 and 5.3. Hence, from the analysis of Table 5.1 and Figures 5.2 and 5.3, we can conclude that for a typical value of a transient background magnetic field of 0.2 µT, the possible distance to the area of FM location lies in the range from 100 to 150 m. At the same time, it is assumed that a stationary magnetic disturbance due to inductively magnetized elements of the AUV design can be identified and eliminated during the calibration of the VMS on board the AUV. This becomes possible due to the stability of the model parameters of this disturbance.

TABLE 5.1

Relationships for Estimating FM Monitoring Range

FM Magnetic Moment Estimation		Estimation of FM Magnetic Field (B) from Distance (R)	
Moment M_p from permanent magnetization of FM	Moment M_i from inductive magnetization of FM	$B = M/R^3$ $R > l$	$B = M/l R^2$ $R < l$
$M_p = B_{res} V$ $B_{res} = 10\ \mu T$—residual induction of the material; $V = 7854\ m^3$—FM volume	$M_i = B_n V(m-1)/(m+2)$ $B_n = 50\ \mu T$—normal EMF; $m = 10$—relative magnetic permeability of FM material		$l = 100$ m–length of the cylindrical FM, $D = 10$ m—FM diameter

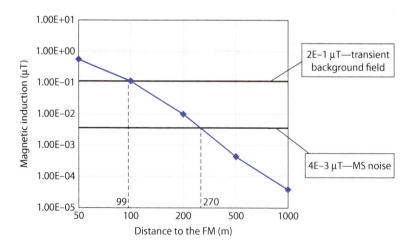

FIGURE 5.2
FM monitoring range and permanent magnetization ($B_{res} = 0.5\ \mu T$).

Application of Magnetic Sensors

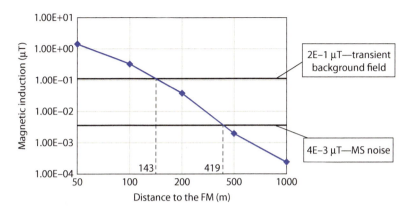

FIGURE 5.3
FM monitoring range and inductive magnetization ($B_{res} = 10 \, \mu T$).

5.2.4 Overview of Typical MS Magnetometers

Upon choosing the VMS for monitoring FMs, it should be taken into account that the entire level of the transient background magnetic field can be up to 0.3 µT, while the total induction of the magnetic field can be up to 100 µT. Key types of MS magnetometers and their characteristics are shown in Table 5.2.

An analysis of Table 5.2 shows that is advisable to use magnetoresistive MS (fabricated by Honeywell) for monitoring FMs from on board the AUV. The operating principle of this type of MS is based on changing the electrical resistance of the ferromagnetic material (magnetoresistor, e.g., permalloy—80% Ni and 20% Fe) subjected to the magnetic field. The sensitivity of magnetoresistive MS to the magnetic field is also dependent on the value of the induction of the magnetic field being measured and on the angle between the vector of induction and the direction of the current flowing through the ferromagnetic material. The sensor provides the calibration

TABLE 5.2

MS Types and Their Typical Parameters

Sensor Type Parameter	Hall Sensors	Inductive Sensors	Fluxgate Sensors	Protonic Sensors	Magnetoresistive Sensors
Measurement range (mT)	From 01 to $3*10^4$	From 10^{-10} to 10^6	From 10^{-4} to 1	From 0.02 to 0.1	From 10^{-3} to 5
Frequency range (Hz)	Up to 10	From 0.1 to 10^6	Up to 2000	Up to 2	Up to 10^7
Resolution (µT)	0.1	10^{-3}	From 10^{-3} to 10^{-6}	From 10^{-4} to 10^{-5}	10^{-2}
Magnetometer type	Vectorial	Vectorial	Vectorial	Scalar	Vectorial
Temperature range (°C)	From −10 to +40	From −4 to +40	From −25 to +55	From −20 to +50	From −55 to +150

mode and compensation for the background magnetic disturbance. In using microtechnology, magnetoresistive MS has a small size and a relatively low cost. Since the MS has only one measuring axis, three sensors are required for measuring the vector of magnetic induction.

Employing an MP, such as a UWTP, for monitoring FMs located on local land surfaces, the inductive MS is used. Because this MS works effectively as a low-frequency electromagnetic field receiver, it is used in an EONS, as will be discussed later. Mounted in the UWTP, the inductive MS is involved in determining the relevant kinematic motion parameters of this MP.

5.3 Calibration of Vector Magnetic Sensors: Magnetometers

As mentioned in Section 5.2, for monitoring an FM from on board an MP, the background magnetic field on board the MP should be restored. For this purpose, the deviation coefficients of the magnetic field model created by inductively magnetized ferromagnetic elements of the MP (so-called soft iron distortion) are defined. Then, the magnetic field created by the permanent magnetization of MP elements (so-called hard iron distortion) and the EMF are also defined.

For further analysis, let us introduce the coordinate frame (CF) fixed to the AUV denoted as $OX_cY_cZ_c$ and the CF for inertial orientation denoted as $OXYZ$. Noting that the OX and OY axes are lying on the horizontal plane, the position of the $OX_cY_cZ_c$ CF with respect to $OXYZ$ is shown in Figure 5.4. The measuring axis of each VMS is directed along one axis of $OX_cY_cZ_c$ CF. The axes of $OXYZ$ CF are modeled by a strapdown inertial orientation

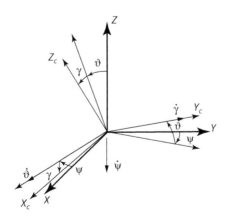

FIGURE 5.4
Coordinate systems.

Application of Magnetic Sensors

system (IOS). The angles corresponding to yaw, pitch, and roll are denoted as ψ, ϑ, γ, respectively.

A model of a magnetic field vector measured by VMSs related to the $OX_cY_cZ_c$ CF can be represented as

$$\mathbf{T_m} = \mathbf{A} \times \mathbf{C} \times \mathbf{T} + \mathbf{T_0} + \mathbf{A_{FM}} \times \mathbf{C} \times \mathbf{T} + \mathbf{T_{FM}} + \mathbf{w}, \tag{5.1}$$

where

$$\mathbf{A} = \begin{bmatrix} a_{xx} & a_{xy} & a_{xz} \\ a_{yx} & a_{yy} & a_{yz} \\ a_{zx} & a_{zy} & a_{zz} \end{bmatrix} \tag{5.2}$$

is the deviation coefficients of the Poisson matrix determining the deformation of the EMF in the MP volume due to "soft iron distortion" [5,6];

$$\mathbf{C} = \begin{bmatrix} \cos\vartheta\cos\psi & \sin\vartheta & -\cos\vartheta\sin\psi \\ \sin\gamma\sin\psi - \cos\gamma\sin\vartheta\cos\psi & \cos\gamma\cos\vartheta & \sin\gamma\cos\psi + \cos\gamma\sin\vartheta\sin\psi \\ \cos\gamma\sin\psi + \sin\gamma\sin\vartheta\cos\psi & -\sin\gamma\cos\vartheta & \cos\gamma\cos\psi - \sin\gamma\sin\vartheta\sin\psi \end{bmatrix} \tag{5.3}$$

is the orientation matrix relative to the *OXYZ* CF [7,8];

$$\mathbf{T} = \begin{bmatrix} T_x T_y T_z \end{bmatrix}^T \tag{5.4}$$

is the EMF vector related to the *OXYZ* CF;

$$\mathbf{T_0} = \begin{bmatrix} T_{0x} T_{0y} T_{0z} \end{bmatrix}^T \tag{5.5}$$

where **w** noise vector is the vector of the MP's own permanent magnetic field related to the $OX_cY_cZ_c$ CF;

$\mathbf{A_{FM}}$ is the deviation coefficients matrix determining the deformation EMF in the FM volume; $\mathbf{T_{FM}}$ is the vector of the permanent magnetic field of the FM; $\mathbf{C_{FM}}$ is the orientation matrix of the FM relative to the *OXYZ* CF; **w** is the vector of noise.

In the absence of an FM, the model of the background magnetic field is rewritten as follows:

$$\mathbf{T_m} = \mathbf{A} \times \mathbf{C} \times \mathbf{T} + \mathbf{T_0} + \mathbf{w}. \tag{5.6}$$

In order to restore this magnetic field, matrix **A** and vectors $\mathbf{T_0}$ and **T** should be estimated. The estimation procedure for **A**, $\mathbf{T_0}$, **T** is called the calibration of the VMS on board the MP.

Let us consider two methods for the calibration of a VMS. The first method is based on the assumption that $\mathbf{A}$, $\mathbf{T}_0$, $\mathbf{T}$ are unchanged during monitoring. Hence, the first method of the calibration is performed onboard the MP before the monitoring starts. The second method is performed during the monitoring process, when $\mathbf{A}$, $\mathbf{T}_0$, $\mathbf{T}$ are assumed to be changing. Next, two common approaches to the implementation of these methods for VMS calibration are considered.

5.3.1 First Method for VMS Calibration

This method is based on the fact that the rotational motion of an MP (AUV) creates a modulation of the projections of the EMF on the axes of $OX_cY_cZ_c$ CF [9,10] (Figure 5.5).

It is assumed that the vector $\mathbf{T}$ is normalized; in other words, that the measurements of T using the model (Equation 5.6) are normalized relative to the EMF. The rotation of the AUV is carried out in such a way as to provide different orientations of vector $\mathbf{T}$ in the $OX_cY_cZ_c$ CS. Rotating a VMS mounted in an AUV, the end point of vector $\mathbf{T}$ describes a smooth ellipsoidal surface in the $OX_cY_cZ_c$ CS, as shown in Figure 5.6.

In the case of an ideal AUV and VMS ($\mathbf{A}$ is the identity matrix, $\mathbf{T}_0$ is the null vector), this surface becomes a sphere centered at the origin of the related CS. The presence of the AUV's own permanent field, $\mathbf{T}_0$ leads to the displacement of the sphere center from the origin of $OX_cY_cZ_c$ CS. The presence of soft iron distortion in the AUV and cross-links between the measuring channels of VMSs lead to the deformation of the sphere (due to the difference of matrix $\mathbf{A}$ from an identity matrix).

Upon rotating the AUV, according to Equation 5.6 the normalized VMS output signal $\mathbf{T}_m / |\mathbf{T}|$ is described by the following relationship:

$$\frac{\mathbf{T}_m}{|\mathbf{T}|} = \mathbf{A} \times \mathbf{C} \times \frac{\mathbf{T}}{|\mathbf{T}|} + \frac{\mathbf{T}_0}{|\mathbf{T}|} + \frac{\mathbf{w}}{|\mathbf{T}|}. \tag{5.7}$$

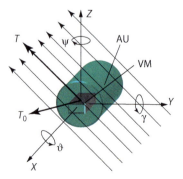

FIGURE 5.5
AUV and VMS in EMF prior to the monitoring process.

Application of Magnetic Sensors

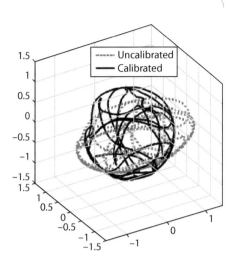

FIGURE 5.6
Surfaces described by the endpoint of vector **T**, upon rotating VMS mounted into an AUV.

Now the VMS calibration is reduced to obtaining the best estimations for components of the A matrix and the T vector, providing the solution to Equation 5.7 for the vector $\mathbf{T}_m/|\mathbf{T}|$. For these estimations, the nonlinear least squares method [11,12] can be used and the following residuals are minimized:

$$\Delta = \sum_{i=1}^{N} \left(|\mathbf{R}_i| - 1\right)^2; \tag{5.8}$$

$$\mathbf{R}_i = \hat{\mathbf{A}}^{-1} \left(\left(\frac{\mathbf{T}_m}{|\mathbf{T}|}\right)_i - \frac{\hat{\mathbf{T}}_0}{|\mathbf{T}|} \right),$$

where:
- $\mathbf{R}_i$ is the radius vector of a point on the surface of the sphere corresponding to the ith measurement of VMSs of the induction of a magnetic field
- $i = 1, 2, 3, \ldots, N$ is the sequence number of the VMS measurement when changing its orientation
- $\hat{\mathbf{A}}$ is the matrix **A** estimation
- $\hat{\mathbf{T}}_0$ is the vector $\mathbf{T}_0$ estimation

The feature of this VMS calibration is the simplicity of its technical implementation. This is because this VMS calibration is performed autonomously and doesn't require information about the AUV orientation.

5.3.2 Second Method for VMS Calibration

The core of the calibration restoring the background magnetic field on board the AUV in the area of FM monitoring is as follows [13,14]. As in Section 5.3.1, model 5.6 is considered the model of a background magnetic field. During the FM monitoring process, the magnetic field on board the AUV is measured by means of a VMS in the CF related to the AUV. At the same time, the AUV angular orientation is defined by, for instance, IOS. Based on this information, the model parameters are estimated, and the vector of induction of the background magnetic field on board the AUV is restored on a trajectory of its motion.

The analysis shows that the accuracy of the estimation of the model parameters of the background magnetic field is mostly dependent on an algorithm, as well as on a profile of changing the AUV angular orientation on its motion trajectory.

5.3.2.1 Algorithm for Parameters Estimation of Background Magnetic Field Model

Using Equations 5.2 through 5.6, the measurements in each VMS channel can be presented in the following form:

$$T_{mx} = c_{11} a_{xx} T_x + c_{12} a_{xx} T_y + c_{13} a_{xx} T_z + c_{21} a_{xy} T_x + c_{22} a_{xy} T_y + c_{23} a_{xy} T_z + c_{31} a_{xz}$$

$$T_x + c_{32} a_{xz} T_y + c_{33} a_{xz} T_z + T_{ox} + w_x = \begin{bmatrix} 1 \; c_{11} \; c_{12} \; c_{13} \; c_{21} \; c_{22} \; c_{23} \; c_{31} \; c_{32} \; c_{33} \end{bmatrix} \quad (5.9)$$

$$\begin{bmatrix} T_{ox} \; Xx_1 \; Xx_2 \; Xx_3 \; Xx_4 \; Xx_5 \; Xx_6 \; Xx_7 \; Xx_8 \; Xx_9 \end{bmatrix}^T + w_x = H \, X_x + w_x;$$

$$T_{my} = c_{11} a_{yx} T_x + c_{12} a_{yx} T_y + c_{13} a_{yx} T_z + c_{21} a_{yy} T_x + c_{22} a_{yy} T_y + c_{23} a_{yy} T_z +$$

$$\begin{bmatrix} c_{31} a_{yz} T_x \end{bmatrix} + c_{32} a_{yz} T_y + c_{33} a_{yz} T_z + T_{oy} + w_y = \begin{bmatrix} 1 \; c_{11} \; c_{12} \; c_{13} \; c_{21} \; c_{22} \; c_{23} \; c_{31} \; c_{32} \; c_{33} \end{bmatrix}$$

$$\begin{bmatrix} T_{oy} \; Xy_1 \; Xy_2 \; Xy_3 \; Xy_4 \; Xy_5 \; Xy_6 \; Xy_7 \; Xy_8 \; Xy_9 \end{bmatrix}^T + w_y = H \, X_y + w_y;$$

$$T_{mz} = c_{11} a_{zx} T_x + c_{12} a_{zx} T_y + c_{13} a_{zx} T_z + c_{21} a_{zy} T_x + c_{22} a_{zy} T_y + c_{23} a_{zy} T_z +$$

$$\begin{bmatrix} c_{31} a_{zz} T_x \end{bmatrix} + c_{32} a_{zz} T_y + c_{33} a_{zz} T_z + T_{oz} + w_z$$

$$= \begin{bmatrix} 1 \, c_{11} \, c_{12} \, c_{13} \, c_{21} \, c_{22} \, c_{23} \, c_{31} \, c_{32} \, c_{33} \end{bmatrix} \begin{bmatrix} T_{oz} \; Xz_1 \; Xz_2 \; Xz_3 \; Xz_4 \; Xz_5 \; Xz_6 \; Xz_7 \; Xz_8 \; Xz_9 \end{bmatrix}^T$$

$$+ w_z = H \, X_z + w_z,$$

where:
c_{ij} ($i, j = 1, 2, 3$) are the elements of orientation in matrix 5.3
a_{nm} ($n, m = x, y, z$) are the elements of the Poisson matrix 5.2

Application of Magnetic Sensors

$$H = \begin{bmatrix} 1 & c_{11} & c_{12} & c_{13} & c_{21} & c_{22} & c_{23} & c_{31} & c_{32} & c_{33} \end{bmatrix};$$

$$X_x = \begin{bmatrix} T_{ox} & Xx_1 & Xx_2 & Xx_3 & Xx_4 & Xx_5 & Xx_6 & Xx_7 & Xx_8 & Xx_9 \end{bmatrix}^T;$$

$$X_y = \begin{bmatrix} T_{oy} & Xy_1 & Xy_2 & Xy_3 & Xy_4 & Xy_5 & Xy_6 & Xy_7 & Xy_8 & Xy_9 \end{bmatrix}^T; \quad (5.10)$$

$$X_z = \begin{bmatrix} T_{oz} & Xz_1 & Xz_2 & Xz_3 & Xz_4 & Xz_5 & Xz_6 & Xz_7 & Xz_8 & Xz_9 \end{bmatrix}^T;$$

T_{mx}, T_{my}, T_{mz} are the components of the measured vector of induction of the magnetic field in relation to CS.

The components of the state vectors X_x, X_y, X_z are the components of the product of the matrix $\mathbf{A}$ and the vector of the magnetic field, $\mathbf{T}$. They are subject to the estimation as model parameters of a background magnetic field onboard the AUV based on solving Equation 5.9. It is easy to see that there are 15 linear equations defining the linear relationships between these components. Basically, the 30th order of set Equation 5.9 can be reduced to the 15th order. However, the analysis of numerical and experimental results shows that from a practical point of view the parameters of Equation 5.10 based on set Equation 5.9 should be estimated in the presence of the real noise $\mathbf{w}$.

In this case, the efficient algorithm is one that is based on the method of least squares (MLS) applied to each of the measurement channels individually and described by the following equation:

$$\hat{X}_i = \left(H^T \times H \right)^{-1} \times H^T \times T_{mi}, \quad i = x, y, z, \quad (5.11)$$

where $\hat{X}_i$ is the estimation for the vector of model parameters X_i, including (each channel) the vector projection of induction of the permanent magnetic field of AUV, T_0, on the respective axes and nine components of the vectors X_x, X_y, X_z and EMF projections on the axes of the *OXYZ* CS.

It is worth noting that upon employing the set Equation 5.9 instead of Equation 5.11 for monitoring an FM from onboard the AUV in real time, the quasi-recurrent algorithm of MLS should be used.

5.3.2.2 Optimization Algorithm for Changing AUV Angular Orientation on Trajectory of Its Motion

In the framework of the admissible range in the variation of AUV orientation angles (Ψ, ϑ, γ), the accuracy of the parameters estimation of model 5.9 depends essentially on the behavior of the process changing these angles with time on a trajectory of the AUV motion. The optimization of this process can be solved based on a D-optimal experimental design [8].

Hence, using Equation 5.9, equations for a single channel of VMS are as follows:

$$T_{mi} = H(\psi, \upsilon, \gamma) X_i + w_i, \quad i = x, y, z. \tag{5.12}$$

Let us introduce the dispersion matrix of errors, **D**, for the vector $\mathbf{X}_i$ estimation and the information matrix, **M**(y), which are as follows:

$$\mathbf{D} = \mathbf{M}^{-1}(y)\sigma^2, \quad \mathbf{M}(y) = \mathbf{H}(y)^T \mathbf{H}(y), \tag{5.13}$$

where:

$$\mathbf{H}(y) = \left[\mathbf{h}_1^T(y) \ldots \mathbf{h}_N^T(y) \right]^T$$

- *y*   is the vector of the optimization parameters (in our case y^T=[ψ, υ, γ])
- σ   is the standard deviation of error measurement

In this case, the optimization algorithm building the optimal profile of changing orientation angles, **y** with time is reduced to minimize the determinant value of the dispersion matrix **D** (or to maximize the determinant value of the information matrix, **M**).

A design set is a collection of three-dimensional space points with the ψ, υ, γ coordinates. Since the most constructive results of the experimental design theory were obtained for a one-dimensional collection [15,16], the mapping of a 3-D dimension onto 1-D based on the Peano space-filling curves is used [17,18]. The continuous design is as follows:

$$\xi = \begin{Bmatrix} y_1 & , \ldots, & y_n \\ p_1 & , \ldots, & p_n \end{Bmatrix}, p_i \geq 0, \sum_{i=1}^{n} p_i = 1, \tag{5.14}$$

where:

- y_i ($i = 1, \ldots, n$) is the design points (i.e., changing the angles within a range of ±20° and with a sampling step of 2°, we have 21 points for each variable and $n = 21^3 = 9261$ design points in total)
- p_i is the weight of the design points, proportional to a number of measurements at the point of optimal design, y_i

The algorithm for finding the optimal design is built for the functional $\Phi[D] = \ln(\det D)$ (generalized D-optimality); it includes the following steps (algorithm).

Step 1. Choose a non-degenerate initial design, ξ_s, type of Equation 5.14 at step number $s = 0$.

Step 2. Search for a design with unit measure, for the rewritten point y_s:

$$y_s = \mathrm{argmax}\big(d(y,\xi_s)\big), \gamma$$

where $d(y,\xi_s) = \mathbf{h}^T(y)\mathbf{D}(\xi_s)\mathbf{h}(y)$.

Step 3. Define the step length γ_s:

$$\gamma_s = \big[d(y,\xi_s)-m\big]\big/\big[(d(y,\xi_s)-1)m\big],$$

where m is the size of $\mathbf{M}$ matrix.

Step 4. Build the design:

$$\xi_s = (1-\gamma_s)\xi_s + \gamma_s \xi(y_s)$$

and then return to Step 2, replacing s with $s+1$.

The iteration procedure described is completed when an asymptotically decreasing increment of the functional $\Phi[D]$ is reached with an acceptable low increment value.

It is worth noting that the step length specified in Step 3 of the algorithm provides a high convergence rate and is obtained from the following equation:

$$\gamma_s = \mathrm{argmin}\big[\mathbf{D}\big((1-\gamma)\xi_s + \gamma\xi(y_s)\big)\big], \gamma$$

At the same time, the dispersion matrix on each step is defined as follows:

$$\mathbf{D}(\xi_{s+1}) = \frac{1}{1-\gamma}\left[\mathbf{E}_m - \frac{\gamma \mathbf{D}(\xi_s)h(y)h^T(y)}{1-\gamma+\gamma d(y,\xi_s)}\right]\mathbf{D}(\xi_s),$$

where $\mathbf{E}_m$ is the identity matrix.

To demonstrate the algorithm's operability, let us consider an example of the program written in the MATLAB environment. The initial distribution of the weights p_i for 9261 design points was chosen either homogeneously or by arbitrary sampling followed by normalization. In both cases, there was a convergence to a single optimal design. The resulting optimal design in the 3-D space of the orientation angles (the limit value of AUV orientation angles was set to 20°) at a sufficient number of iterations equal to 500, comprises 22 points and is shown in Figure 5.7 and Table 5.3. The information matrix $\mathbf{M}$ determinant value of the normalized optimal design is 9.1074e-19, and the initial (evenly distributed over all points) design is 2.7042e-22.

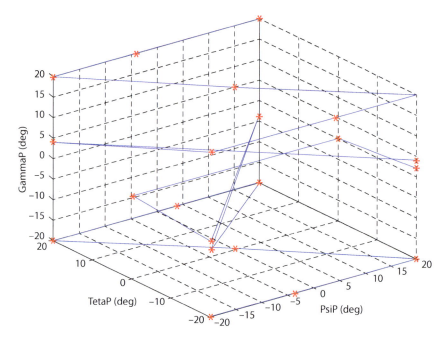

FIGURE 5.7
Dependency of GammaP from PsiP, TetaP.

The points of the optimal design are connected according to their numbering in Table 5.3 and as shown in Figure 5.7. The first row of Table 5.3 specifies the point number from the general design. The second row specifies the normalized weight of this point of the design. The third, fourth, and fifth rows specify the corresponding values of the yaw, pitch, and roll angles, respectively. Guidance on changing the angular orientation of the AUV in time can be performed by using the values of the design points from Table 5.3 (excluding the reorientation costs). An example of the reorientation guidance is shown in Figure 5.8. In these figures, the abscissa axis is dimensionless time t/T (t is the time spent by the object at the corresponding design point, proportional to the weight of this point p, T is the total time of the experiment). If we consider the reorientation costs of the AUV, in particular the energy costs, and also the dynamic features of the AUV, then in order to build an improved reorientation guidance, it is necessary to solve the problem of resources optimization. Thus, in order to minimize reorientation costs, a route bypassing the points of the obtained optimal design should be built. This case is represented by the traveling salesman problem [19,20].

In order to determine the comparative effectiveness of the optimal design for the estimation accuracy of the state vector components, a numerical experiment was performed. This experiment includes calculating and comparing the standard deviations (SD) of estimation errors of the state vector **X**

Application of Magnetic Sensors

TABLE 5.3
Optimal Design Distribution over All Points

No.	1	2	3	4	5	6
Weight	0.089092	0.0048485	0.093241	0.049254	0.093971	0.0036797
Ψ (deg)	−20	−4	20	0	−20	4
υ (deg)	−20	−20	−20	0	20	20
γ (deg)	−20	−20	−20	−20	−20	−20
No.	7	8	9	10	11	12
Weight	0.091295	0.0033493	0.0038517	0.002693	0.04865	0.050248
Ψ (deg)	20	−20	20	−20	−20	20
υ (deg)	20	−20	20	−20	0	0
γ (deg)	−20	−4	−4	−2	0	0
No.	13	14	15	16	17	18
Weight	0.0027813	0.00176	0.0040216	0.094233	0.0048343	0.09131
Ψ (deg)	20	20	−20	−20	4	20
υ (deg)	−20	−20	20	−20	−20	−20
γ (deg)	2	4	4	20	20	20
No.	19	20	21	22		
Weight	0.049492	0.090595	0.0039026	0.094668		
Ψ (deg)	0	−20	−4	20		
υ (deg)	0	20	20	20		
γ (deg)	20	20	20	20		

by MLS in optimal and non-optimal designs. The SD of estimation errors $\sigma_i()$ is defined by the following equation:

$$\sigma_i^2() = \left[H(\xi_i)^T H(\xi_i) \right]^{-1} \sigma^2, \quad i=1,2,$$

where:

- ξ_1 is the optimal design of the experiment
- ξ_2 is the experiment design based on the use of standard evolutions of AUV ("snake movement" for yaw angle, "climbing movement" for pitch angle, and "rolling movement" for roll angle, Figure 5.9)
- σ is the SD of the measurement errors (0.1 μT)

The calculation results are shown in Table 5.4.

An analysis of Table 5.4 shows that the SD for the state vector estimations obtained in the optimal experiment design are smaller than the SD of state vector estimations obtained in the non-optimal experiment design, ranging from 3 to 20 times.

Noting that the analytical solution of a similar problem, for building an optimal design for the AUV roll angle, applied to the evaluation of drifts of the gyroscopes using IOS was obtained in [21,22]. This analytical solution

150 Magnetic Sensors and Devices

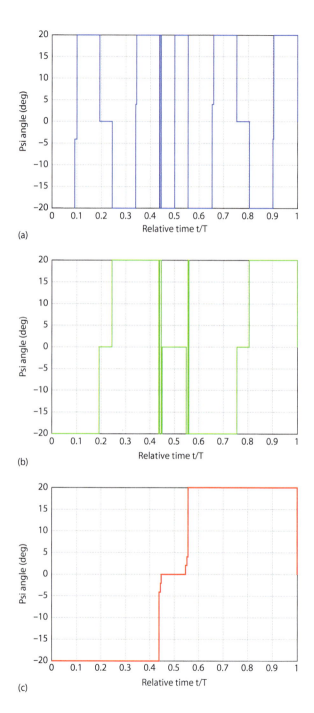

FIGURE 5.8
(a) Heading angle optimal design and time. (b) Pitch angle optimal design and time. (c) Roll angle optimal design and time.

Application of Magnetic Sensors

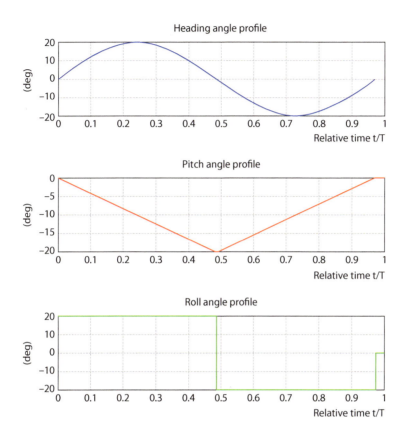

FIGURE 5.9
Attitude angles profile: Standard approach.

provides the same results as the results obtained in the developed algorithm for the 1-D case of the roll angle.

The obtained results substantiate the possibility of optimizing the process of determining the induction vector of the background magnetic field on board the AUV by directional variation of its angular orientation on a trajectory of its motion.

5.4 Determining Location of FM Relative to Mobile Platform and Its Transportation to Area of FMs Location

5.4.1 Determining Location of FM

Let us assume that the distance to the area of the FM location is sufficiently large. Then, the magnetic field induced by the FM can be adequately modeled

TABLE 5.4
SD of State Vector's **X** Estimation Errors

	$\sigma(X_1)$, µT	$\sigma(X_2)$, µT	$\sigma(X_3)$, µT	$\sigma(X_4)$, µT	$\sigma(X_5)$, µT	$\sigma(X_6)$, µT	$\sigma(X_7)$, µT	$\sigma(X_8)$, µT	$\sigma(X_9)$, µT	$\sigma(X_{10})$, µT
Design ξ_1	0.0274	0.0399	0.0097	0.0114	0.0103	0.0399	0.0108	0.0108	0.0114	0.0290
Design ξ_2	0.4685	0.3935	0.2131	0.0582	0.1091	0.4648	0.1167	0.0380	0.1255	0.4036

Application of Magnetic Sensors 153

by the field of magnetic dipole. Hence, the objective is to identify components of the radius vector defined in *OXYZ* CS from the FM dipole center to the AUV one. In other words, the problem is reduced to determining the location of the FM relative to the AUV. It is called the *inverse dipole localization problem*. In the known methods of its solution, measured values of the vector of magnetic induction induced by the FM, and the absolute value of its magnetic induction, and their gradients are used as the original information [23,24]. Thus, to achieve the desired result on board the AUV, a magnetometer and a magnetic gradiometer are required. Building a magnetic gradiometer based on combining data from several magnetometers on board the AUV is a difficult technical problem. The main difficulties are as follows:

1. The design of an AUV does not allow setting an extended and stable base between the magnetometer and the gradiometer due to the small size of the AUV, the high "packing" density of its units and systems, and the stiffness of the AUV design, which is not high enough.
2. Hence, a short base is used. As a result, for realizing a gradiometer on short base magnetometers with a high sensitivity, a large dynamic range, a high linearity, and stability of its characteristics are required. Such magnetometers have a high cost, large size, and low resistance to perturbing operational factors.
3. The identity of the characteristics of the magnetometers under a high level of magnetic disturbance generated by equipment mounted in the AUV needs to be obtained.
4. The calibration during the FM monitoring process significantly increases the difficulty of compensating for magnetic disturbance due to the magnetic gradients of the AUV's own field. Intensive filtering of this magnetic disturbance is required (at the level of 120 dB).

Due to the foregoing difficulties, using one VMS to measure the magnetic gradient is the effective approach to solving the problem of determining the FM location relative to the AUV. The gradient is calculated by measuring the increment of the vector of the magnetic induction upon moving the AUV along the known trajectory. The parameters of this known trajectory are defined by an onboard navigation system, for instance, a strapdown inertial navigation system (INS) [7,8].

5.4.2 Algorithms for Determining FM Location Relative to AUV

The vector of the dipole magnetic induction is defined as follows [23,24]:

$$\mathbf{B} = \frac{\mu_0}{4\pi} \left[\frac{3(\mathbf{M} \cdot \mathbf{R}) \cdot \mathbf{R}}{R^5} - \frac{\mathbf{M}}{R^3} \right], \qquad (5.15)$$

where:
- μ_0 is the magnetic constant
- **R** is the radius vector from the dipole center to the observation point
- **M** is the vector of the magnetic dipole moment

The gradient of the magnetic field vector is a second-order tensor:

$$\mathbf{D} = \begin{bmatrix} \dfrac{\partial B_x}{\partial R_x} & \dfrac{\partial B_x}{\partial R_y} & \dfrac{\partial B_x}{\partial R_z} \\ \dfrac{\partial B_y}{\partial R_x} & \dfrac{\partial B_y}{\partial R_y} & \dfrac{\partial B_y}{\partial R_z} \\ \dfrac{\partial B_z}{\partial R_x} & \dfrac{\partial B_z}{\partial R_y} & \dfrac{\partial B_z}{\partial R_z} \end{bmatrix}, \quad (5.16)$$

where x,y,z are the indexes indicating the projections of the vector on the axes of **OXYZ** CS, respectively.

There are several methods for solving the inverse dipole localization problem.

Method 1:

$$\mathbf{D} \times \mathbf{R} = -3\mathbf{B};$$
$$\mathbf{R} = -3\mathbf{D}^{-1}\mathbf{B}, \quad (5.17)$$

Method 2:

$$\mathbf{R} = -\frac{3}{\lambda_2 \lambda_3}(\mathbf{D} + \lambda_1 \mathbf{I}) \cdot \mathbf{B}, \quad (5.18)$$

Method 3:

$$\mathbf{R} = -3\mathbf{D}^{-2} B \cdot \nabla B, \quad (5.19)$$

Method 4:

$$\mathbf{R} = \frac{3}{\lambda_2 \lambda_3}(B \cdot \nabla B + \lambda_1 \mathbf{B}), \quad (5.20)$$

where:
- $\lambda_1, \lambda_2, \lambda_3$ are the eigenvalues of the characteristic equation of tensor **D**
- **I** is the identity matrix

$$B = c \cdot \sqrt{\frac{3 \cdot (\mathbf{MR})^2}{R^8} + \frac{M^2}{R^6}}$$

$$\nabla B = \frac{\partial B}{\partial x}\mathbf{i} + \frac{\partial B}{\partial y}\mathbf{j} + \frac{\partial B}{\partial z}\mathbf{k} = \mathbf{G} = G_x\mathbf{i} + G_y\mathbf{j} + G_z\mathbf{k}$$

$$G_x = \frac{c^2 \cdot \left(6 \cdot M_x \cdot (\mathbf{MR}) \cdot R^2 - 24 \cdot R_x \cdot (\mathbf{MR})^2 - 6 \cdot M^2 \cdot R^2 \cdot R_x\right)}{2 \cdot B \cdot R^{10}} \quad (5.21)$$

$$G_y = \frac{c^2 \cdot \left(6 \cdot M_y \cdot (\mathbf{MR}) \cdot R^2 - 24 \cdot R_y \cdot (\mathbf{MR})^2 - 6 \cdot M^2 \cdot R^2 \cdot R_y\right)}{2 \cdot B \cdot R^{10}}$$

$$G_z = \frac{c^2 \cdot \left(6 \cdot M_z \cdot (\mathbf{MR}) \cdot R^2 - 24 \cdot R_z \cdot (\mathbf{MR})^2 - 6 \cdot M^2 \cdot R^2 \cdot R_z\right)}{2 \cdot B \cdot R^{10}}$$

B   is the magnitude of the vector of magnetic field induction
∇B   is the gradient of the magnitude of the magnetic field induction

$$c = \frac{\mu_0}{4\pi}$$

Thus, in the framework of the foregoing methods, solving the inverse dipole localization problem based on using one VMS can be reduced to the following algorithm [25,26]:

1. Magnetic induction is measured at two different points on a trajectory of the AUV motion. Also, the coordinates of these points are determined by a strapdown INS.
2. The increments for each component of the magnetic induction vector and coordinates are calculated.
3. Using these increments, partial derivatives from Equations 5.16 through 5.21 are calculated. At the same time, it is assumed that the gradient of the magnetic field between the points, in which the measurements occur, is constant.
4. Employing one from the following Equations 5.17 through 5.20, the position of the dipole relative to the AUV (components of the radius vector **R** from the FM dipole center to the AUV one) is calculated.

5.4.3 Transportation of MP to the Area of FM Location

In a general case, the magnetic "image" of an FM is an image of random type. Moreover, the corresponding model of the FM magnetic field is significantly different from the dipole model in the nearest area of the FM location. Hence, the abovementioned approach to determining the FM location relative to the AUV is inadequate for the nearest area of its location. Therefore, the transportation of the MP to the FM location is solved by the guidance of the MP, which is independent of the shape of the magnetic field induced by the FM, based on the measured and calculated parameters of the FM magnetic field providing the motion of the MP on a trajectory leading to the location. For such guidance of the MP, it is reasonable to use the vector of gradient of magnetic field magnitude (Equation 5.21), which is calculated by means of the measured increment of magnitude of induction of the magnetic field on a trajectory of the AUV motion [27,28]. The guidance is carried out in such a way that in each moment of time, the direction of the longitudinal AUV axis coincides with the current direction of the vector gradient of the FM magnetic field magnitude. In a general case, this direction does not have the same direction as the radius vector from the FM dipole center to the AUV center. The closer the AUV comes to the FM, the more the vector of magnitude's gradient is directed toward the FM. Simultaneously, the dependence of this magnitude on the distance to the FM is a potential and smooth function even for the complex image of an FM magnetic field.

5.5 Electromagnetic System for Positioning and Azimuth Angular Orientation of Ground Mobile Platform

Monitoring an FM located on local land surfaces can be realized based on previously discussed methods in Sections 5.2 and 5.3, using a UWTP as the ground MP. The design of a UWTP and its operating principle are similar to a robotic vehicle such as a Segway [2]. A UWTP is a uniaxial, two-wheeled vehicle with its own platform for measurement equipment. The platform has a rotational degree of freedom around the axis of the wheels, relative to the horizon plane (α angle). Counter-rotating wheels provide the azimuth rotation of the UWTP, characterized by angle ψ. The non-horizontality of the surface under the UWTP is characterized by the roll angle, γ. The position of the UWTP on the local land surface is evaluated using a reference point.

The positioning and orientation in particular around the azimuth angle ψ can be implemented by an EONS. The advantages of an EONS are its simple design and small size.

Application of Magnetic Sensors

5.5.1 EONS Configuration

An EONS consists of a ground-based emitter setting at the reference point and an onboard receiver setting in the UWTP. The scheme of an EONS is shown in Figure 5.10. The ground-based emitter of EONS with frame antennas includes a low-frequency controlled generator and an emission mode control unit with a sync signal transmitter. The radio channel (RC) provides the synchronization of the emitted and received signals.

The onboard equipment of a UWTP includes a magnetic field receiver (inductive vector magnetometer), a receiver of sync signals, and a signal processing unit (SPU). In turn, the SPU provides orientation and navigation of the UWTP, eliminates errors, and generates output signals.

The relative position of an EONS emitter and receiver is shown in Figure 5.11. In this figure, $AX_E Y_E Z_E$ is the fixed CF related to the emitter (the point A is the reference one, the X_E and Y_E axes are lying on the horizon plane, the AY_E axis determines a basic direction), and $OX_R Y_R Z_R$ is the CS related to the EMSON receiver (the point O is assigned to the center of the UWTP platform). The distance between CF centers is characterized by the radius vector $\mathbf{R}$, see Figure 5.11a.

Three angles, namely, the yaw angle, ψ, the inclination angle of the platform, α, and the roll angle, γ, characterize the angular position of the $X_R Y_R Z_R$ CS with respect to the $X_E Y_E Z_E$ CS as shown in Figure 5.11b.

The relative orientation of the listed CSs is determined by the direction cosine matrix $A_{R/E}$, which is as follows:

$$A_{R/E} = \begin{bmatrix} \cos\alpha\cos\psi & \cos\psi\sin\alpha\sin\gamma - \sin\psi\cos\gamma & \cos\psi\sin\alpha\cos\gamma + \sin\psi\sin\gamma \\ \cos\alpha\sin\psi & \sin\psi\sin\alpha\sin\gamma + \cos\psi\cos\gamma & \sin\psi\sin\alpha\cos\gamma - \cos\psi\sin\gamma \\ -\sin\alpha & \cos\alpha\sin\gamma & \cos\alpha\cos\gamma \end{bmatrix}.$$

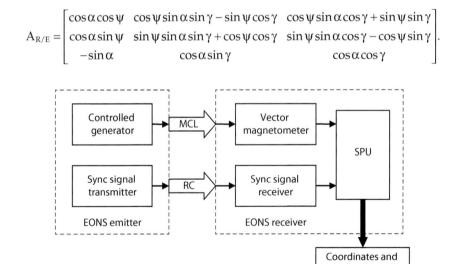

FIGURE 5.10
Scheme of EONS: MCL is the magnetic communication line; RC is the radio channel; SPU is the signal processing unit.

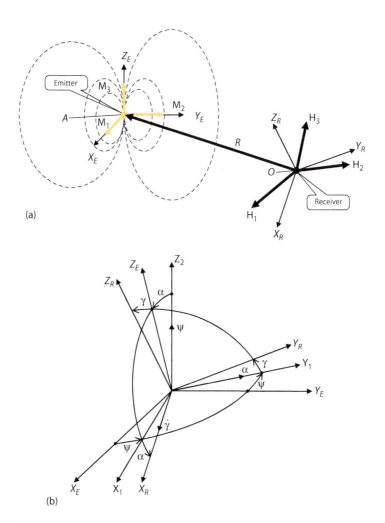

FIGURE 5.11
Relative position of an EONS emitter and receiver CS.

The matrix $A_{R/E}$ maps the projections of a vector from $X_R Y_R Z_R$ CS into $X_E Y_E Z_E$ CS. In this case, the angle of the UWTP azimuth orientation relative to the basic direction is defined as

$$\psi = \text{arctg}\left(\frac{a_{21}}{a_{11}}\right), \tag{5.22}$$

where a_{21}, a_{11} are the corresponding elements of matrix $A_{R/E}$. In a similar way, the angles α and γ can also be defined in terms of elements of matrix $A_{R/E}$.

The power supply of mutually perpendicular frame antennas of the emitter is performed sequentially, and provided by a controlled generator. The timing signal provides the sequential connection of the controlled generator

Application of Magnetic Sensors

to the respective frame antennas. At the same time, the index of the currently emitting antenna is reported to the controlled generator through the RC receiver.

The vector magnetometer measures the three components of the navigation magnetic field emitting frame antennas. A signal diagram for the EONS channel is shown in Figure 5.12. In this diagram, M_i, H_i ($i = 1,2,3$) are the magnetic moments generated by the ith antenna and the magnetic field intensity at the point O (observation point) generated by these moments, respectively; H_{ij} ($i = 1,2,3$; $j = X_R Y_R Z_R$) are the components of the magnetic field intensity, created in every jth receiver's magnetometer by the ith emitter's antenna.

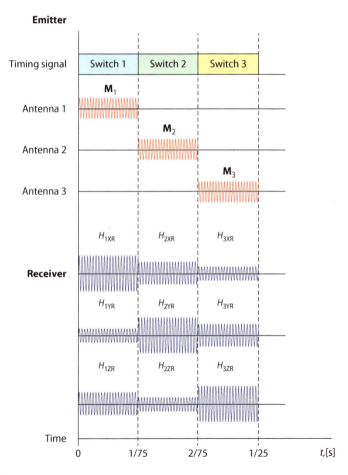

FIGURE 5.12
Signal diagram for EONS channel.

According to the magnetometer measurements, the matrix of signals **H** is formed in an SPU. The elements of this matrix are proportional to the projections of the magnetic field's intensity on the axis of $X_R Y_R Z_R$ CS:

$$\mathbf{H} = \begin{bmatrix} H_{1X_R} & H_{1Y_R} & H_{1Z_R} \\ H_{2X_R} & H_{2Y_R} & H_{2Z_R} \\ H_{3X_R} & H_{3Y_R} & H_{3Z_R} \end{bmatrix}. \tag{5.23}$$

5.5.2 Orientation and Positioning by Means of EONS

The orientation and positioning parameters of an EONS receiver relative to the emitter are determined based on the measurement of the navigation magnetic field at the point O. Estimation of the position and azimuth orientation of the UWTP is performed at a distance from the emitter larger than the linear dimensions of the emitter antennas of more than one order. Hence, let us use the dipole representation of the navigation magnetic field intensity at the point O.

The value of magnetic moment, $\mathbf{M}_i$, created by the corresponding emitter antenna is as follows [15]:

$$\mathbf{M}_i = H_i R^3 \left(\frac{3 \; \mathbf{R} \; \mathbf{RH}_i}{2 \; R \; RH_i} - \frac{\mathbf{H}_i}{H_i} \right), \quad i = 1, 2, 3. \tag{5.24}$$

The scalar product of magnetic moments, for example, $\mathbf{M}_1 \mathbf{M}_2$, created by the first antenna and the second one of the emitter (note that the antennas are mutually perpendicular), is as follows:

$$\mathbf{M}_1 \mathbf{M}_2 = H_1 R^3 \left(\frac{3 \; \mathbf{R} \; \mathbf{RH}_1}{2 \; R \; RH_1} - \frac{\mathbf{H}_1}{H_1} \right) H_2 R^3 \left(\frac{3 \; \mathbf{R} \; \mathbf{RH}_2}{2 \; R \; RH_2} - \frac{\mathbf{H}_2}{H_2} \right) = 0. \tag{5.25}$$

where:

$$\frac{\mathbf{RH}_1}{RH_1} = \cos \alpha_1$$

$$\frac{\mathbf{RH}_2}{RH_2} = \cos \alpha_2$$

$$\frac{\mathbf{R}}{R} = \mathbf{r}$$

$$\frac{\mathbf{H}_1}{H_1} = \mathbf{h}_1$$

$$\frac{\mathbf{H}_2}{H_2} = \mathbf{h}_2$$

Application of Magnetic Sensors

α_1 is the angle between vectors **R** and **H**$_1$
α_2 is the angle between vectors **R** and **H**$_2$
r, **h**$_1$, **h**$_2$ are the basis vectors of **R**, **H**$_1$, **H**$_2$ respectively

Accounting for $\mathbf{h}_1\mathbf{r} = \cos\alpha_1$, $\mathbf{h}_2\mathbf{r} = \cos\alpha_2$, $\mathbf{H}_1\mathbf{H}_2/H_1H_2 = \mathbf{h}_1\mathbf{h}_2$, from condition 5.25, we have $\mathbf{H}_1\mathbf{H}_2/H_1H_2 = \frac{3}{4}\cos\alpha_1\cos\alpha_2$ or

$$\beta_{12} = \tfrac{3}{4}\beta_1\beta_2, \tag{5.26}$$

where:
$\beta_1 = \cos\alpha_1$
$\beta_2 = \cos\alpha_2$
$\beta_{12} = \dfrac{\mathbf{H}_1\mathbf{H}_2}{H_1H_2}$ is the cosine of the angle between vectors **H**$_1$ and **H**$_2$

$$\mathbf{H}_1\mathbf{H}_2 = H_{1X_R}H_{2X_R} + H_{1Y_R}H_{2Y_R} + H_{1Z_R}H_{2Z_R}$$

Writing scalar products of the magnetic moments of other antennas (**M**$_1$**M**$_3$, **M**$_2$**M**$_3$), we obtain a set of three equations for β_1, β_2, β_3, which are cosines of the angles between the radius vector **R** and vectors **H**$_1$, **H**$_2$, **H**$_3$, as follows:

$$\beta_{12} = \tfrac{3}{4}\beta_1\beta_2; \quad \beta_{13} = \tfrac{3}{4}\beta_1\beta_3; \quad \beta_{23} = \tfrac{3}{4}\beta_2\beta_3. \tag{5.27}$$

The solution of set 5.27 becomes

$$\beta_1^2 = \frac{4}{3}\frac{\beta_{12}\beta_{13}}{\beta_{23}}; \quad \beta_2^2 = \frac{4}{3}\frac{\beta_{12}\beta_{23}}{\beta_{13}}; \quad \beta_3^2 = \frac{4}{3}\frac{\beta_{13}\beta_{23}}{\beta_{12}}. \tag{5.28}$$

Directional cosines r_x, r_y, r_z of the radius vector **R** in relation to $X_R Y_R Z_R$ CF are determined by the following equations:

$$\mathbf{h}_1\mathbf{r} = h_{11}r_x + h_{12}r_y + h_{13}r_z = \beta_1;$$
$$\mathbf{h}_2\mathbf{r} = h_{21}r_x + h_{22}r_y + h_{23}r_z = \beta_2; \tag{5.29}$$
$$\mathbf{h}_3\mathbf{r} = h_{31}r_x + h_{32}r_y + h_{33}r_z = \beta_3,$$

where:

$$r_x = \frac{R_x}{R}$$

$$r_y = \frac{R_y}{R}$$

$$r_z = \frac{R_z}{R}$$

h_{ij} ($i, j = 1,2,3$) are the projections of h_i on the axes of $X_R Y_R Z_R$ CF

Knowing the magnitude of vector **R** and the directional cosines r_x, r_y, r_z, the coordinates of the radius vector **R** in relation to the receiver $X_R Y_R Z_R$ CF are calculated as follows:

$$R_x = R r_x; \quad R_y = R r_y \quad R_z = R r_z. \tag{5.30}$$

The angular orientation in relation to the receiver CS ($X_R Y_R Z_R$) relative to the emitter CS ($X_E Y_E Z_E$) can be determined by directional cosines matrix $\mathbf{A}'_{R/E}$:

$$\mathbf{A}'_{R/E} = \begin{bmatrix} \left(\frac{3}{2}r_x\beta_1 - h_{11}\right)\bigg/\sqrt{1-\frac{3}{4}\beta_1^2} & \left(\frac{3}{2}r_y\beta_1 - h_{12}\right)\bigg/\sqrt{1-\frac{3}{4}\beta_1^2} & \left(\frac{3}{2}r_z\beta_1 - h_{13}\right)\bigg/\sqrt{1-\frac{3}{4}\beta_1^2} \\ \left(\frac{3}{2}r_x\beta_2 - h_{21}\right)\bigg/\sqrt{1-\frac{3}{4}\beta_2^2} & \left(\frac{3}{2}r_y\beta_2 - h_{22}\right)\bigg/\sqrt{1-\frac{3}{4}\beta_2^2} & \left(\frac{3}{2}r_z\beta_2 - h_{23}\right)\bigg/\sqrt{1-\frac{3}{4}\beta_2^2} \\ \left(\frac{3}{2}r_x\beta_3 - h_{31}\right)\bigg/\sqrt{1-\frac{3}{4}\beta_3^2} & \left(\frac{3}{2}r_y\beta_3 - h_{32}\right)\bigg/\sqrt{1-\frac{3}{4}\beta_3^2} & \left(\frac{3}{2}r_z\beta_3 - h_{33}\right)\bigg/\sqrt{1-\frac{3}{4}\beta_3^2} \end{bmatrix} \tag{5.31}$$

where $\mathbf{A}'_{R/E} = \mathbf{A}_{R/E}$.

Moreover, substituting the corresponding elements of the matrix 5.31 into Equation 5.22, the equation for ψ angle is as follows:

$$\psi = \arctan\left(\frac{\left(\frac{3}{2}r_x\beta_2 - h_{21}\right)}{\left(\frac{3}{2}r_x\beta_1 - h_{11}\right)} \cdot \frac{\sqrt{1-\frac{3}{4}\beta_1^2}}{\sqrt{1-\frac{3}{4}\beta_2^2}}\right). \tag{5.32}$$

Thus, Equations 5.30 through 5.32 determine the positioning parameters of the UWTP relative to the reference point A, and its azimuth orientation relative to the basic direction of AY_E.

References

1. Ageev M., Kiselev L., and Matvienko U. *Autonomous Underwater Robots: Systems and Technologies*, Nauka, Moscow, 2005. p. 398.
2. Inzartsev A.E. (ed.) *Underwater Vehicles*, Intech, https://www.intechopen.com/books/underwater_vehicles, Austria, 2009. p. 582.

3. Aleshin B., Chernomorcky A., Plehanov V., Makeinov V., and others. Orientation, navigation and stabilization of uniaxial wheeled vehicles, *M.: MAI*, 2012, p. 271.
4. Segway Advanced Development http://www.segway.com/puma/.
5. Blednov V. Method for determination of angular components of geomagnetic field induction vector on board the moving ferromagnetic carriers, *Geophysics*, Academy of Science Reports, 1995, Vol. 341, No. 2, pp. 251–254.
6. Certenais J. and Periou J.J. Electromagnetic measurements at sea (Thomas Marconi Sonar SAS, Brest Department, France). *Proceedings of the International Conference on Marine Electromagnetics*, London, 1997.
7. Aleshin B., Plehanov V., Chernomorsky A., and others. Orientation and navigation of moving objects: Modern information technologies, *M.: Physmathlit*, 2006, p. 422.
8. Titterton D.H. and Weston J.L. *Strapdown Inertial Navigation Technology*, Peter Peregrinus and IEE, London, 1997.
9. Anchutin S., Kuris E., Maximov V., Plehanov V., and Chernomorsky A. Result of experimental studies of vector magnetometer in micromechanical AHRS, *MNTS XVI Proceedings: Modern Technologies in Problems of Control, Automatics and Information Processing*, Tulsky GU, 2007, pp. 89–90.
10. Crassidis J.L., Lai K.-L., and Harman R.R. Real-time attitude-independent three-axis magnetometer calibration, *Journal of Guidance, Control, and Dynamics*, 2005, Vol. 28, No. 1, pp. 115–120.
11. Trifonov A. Optimization problems and numerical methods for their solution. *M.: Business*, 2003, p. 230.
12. Crassidis J.L. and Junkins, J.L. *Optimal Estimation of Dynamic Systems*, Chapter 5, CRC Press, Boca Raton, FL, 2004.
13. Tihonov V., Chernomosky A., Plehanov V., and Maximov V. Optimization of the process of magnetic field induction vector determination on board the moving ferromagnetic object, *Aerospace Instrumentation*, 2005, No. 4, pp. 30–34.
14. Challa M. and Natanson G, Effects of magnetometer calibration and maneuvers on accuracies of magnetometer-only attitude and rate determination, *Proceedings of the AAS/GSFC 13th International Symposium on Space Flight Dynamics*, Vol. 1, (NASA/CP-1998-206858/VOL1) NASA-Goddard Space Flight Center, Greenbelt, MD, May 1998, pp. 389–401.
15. Ermakov S. and Jiglavsky A. Optimal design of experiments mathematic theory, *M.: Science*, 1987, p. 320.
16. Eriksson L., Johansson E., Kettaneh-Wold N., Wikstrom C., and Wold S. *Design of Experiments: Principles and Applications*, Ume, Learnways AB, 2000.
17. Batishev D. Methods of optimal design, *Radio and Communications*, 1984, p. 242.
18. Cook R.D. and Nachtsheim C.J. A comparison of algorithms for constructing exact D-optimal designs, *Technometrics*, 1980, Vol. 22, No. 3, pp. 315–324.
19. Moiseev N. Optimization methods, *M.: Science*, 1982, p. 310.
20. Bertsimas, D. and Tsitsiklis J. *Introduction to Linear Optimization*, Athena Scientific, Belmont, MA, 1997.
21. Tihonov V. and Chernomorsky A. Autonomous detection of gyroscopes drift in strapdown inertial navigation system, *Aerospace Instrumentation*, 2004, No. 1, pp. 24–27.
22. Johnson C. Adaptive corrective alignment for a carouseling strapdown INS, *1984 American Control Conference*, San Diego, CA, 1984, pp. 1856–1861.

23. Semevsky R., Averkiev V., and Yarotsky V. Special magnetometry, *S.Pb.: Science*, 2002, p. 360.
24. Brennan J.B. and Davis T.M. The influence of the natural environment on MAD operations. Technical Report, USA, 1969, September.
25. Plehanov V., Maximov V., Chernomorsky A., and Shvetskova A. Determination of magnitude gradient of the ferromagnetic object's magnetic field on board the mobile platform, *MNTS XVIII Proceedings: Modern Technologies in Problems of Control, Automatics and Information Processing*, Moscow, Moscow Institute of Radio-electronics and Automatics, 2009, p. 110.
26. McFee J.E. and Das Y. Determination of the parameters of a dipole by measurement of its magnetic field, *IEEE Transactions on Antennas and Propagation*, 1981, AP-29 282–286.
27. Plehanov V., Chernomorsky A., and Shvetskova A. Using information about magnitude gradient of magnetic field of ferromagnetic object for its position determination, *MNTS XIX Proceedings: Modern Technologies in Problems of Control, Automatics and Information Processing*, Moscow, Moscow Institute of Electrical Engineering, 2010, p. 292.
28. Arie S., Lerner B., Salomonski N. et al. Localization and magnetic moment estimation of a ferromagnetic target by simulated annealing, *Measurement Science and Technology*, 2007, Vol. 18, pp. 3451–3457.

6

A Model to Calculate Force Characteristics of a Magnetic Suspension of a Superconducting Sphere

Sergey I. Kuznetsov and Yury M. Urman

CONTENTS

6.1 Magnetic Levitation and Its Applications 166
6.2 Superconductor in Magnetic Field 168
6.3 Solving Field Problem 169
    6.3.1 Secondary Field 169
    6.3.2 Integral Equations for Secondary Field 170
    6.3.3 Solution for Spherical Shape 171
    6.3.4 Total Field 173
6.4 Energy, Force Characteristics, and Dynamics 174
    6.4.1 Energy Integrals 174
    6.4.2 Forces 176
    6.4.3 Stiffness 177
    6.4.4 Dynamics 178
6.5 Suspension on Circular Current Loops 179
    6.5.1 System of Circular Current Loops 179
    6.5.2 One Current Loop 180
6.6 Suspensions on Permanent Magnets 182
    6.6.1 Permanent Magnets and Magnetic Charges 182
    6.6.2 Force Function of a Sphere in a Suspending Field Produced by N Point Magnetic Poles 183
    6.6.3 Two Point Magnets 185
    6.6.4 Circular Magnet 190
6.7 Conclusions 196
References 196

ABSTRACT: In magnetic levitation systems, a magnetized object is held in space by magnetic force interaction and without mechanical contact to a support. The absence of mechanical contact minimizes friction, wear, and energy dissipation, motivating their use in a number of applications[1-3], including motion creation systems (motors and precise positioning for microsystems [4], orientation systems for space vehicles [5]), sensors (accelerometers, gravimeters, inclinometers, gyroscopes, seismometers), motion storage systems (energy and momentum storages [6]), and systems for harvesting energy [7]. Magnetic levitation systems make it possible to isolate an object from external influences of extreme environments, such as mechanical, heat, and chemical, to cancel vibrations [8,9], create microgravity [10–12], providing means for new approaches to material synthesis and processing, such as crystal, protein, bacterial cluster growth [13], precise measurement of thermophysical properties [14], manipulation and separation of small-scale objects [15,16], including bio-objects [17].

6.1 Magnetic Levitation and Its Applications

In magnetic levitation systems, a magnetized object is held in space by magnetic force interaction and without mechanical contact to a support. The absence of mechanical contact minimizes friction, wear, and energy dissipation, motivating their use in a number of applications [1–3], including motion creation systems (motors and precise positioning for microsystems [4], orientation systems for space vehicles [5]), sensors (accelerometers, gravimeters, inclinometers, gyroscopes, seismometers), motion storage systems (energy and momentum storages [6]), and systems for harvesting energy [7]. Magnetic levitation systems make it possible to isolate an object from the external influences of extreme environments, such as mechanical, heat, and chemical influences, to cancel vibrations [8,9] and create microgravity [10–12], providing the means for new approaches to material synthesis and processing, such as crystal, protein, and bacterial cluster growth [13]; the precise measurement of thermophysical properties [14]; and the manipulation and separation of small-scale objects [15,16], including bio-objects [17]. Such systems are natural to a cryogenic environment and require little to no energy supply, making them particularly appealing for aerospace applications, especially for micro- and nano-satellites, where power supply, spatial dimensions, and weight are extremely limited, while high precision measurements of orientation and acceleration are required [18].

Of particular interest are the passive levitation systems that are intrinsically stable and thus do not need control systems, which significantly reduces the complexity and cost of a suspension system and increases its robustness. According to the Earnshaw–Braunbeck theorem, stable passive levitation in free space is only possible when diamagnetic materials are present either on the levitated body or in the supporting system [19,20]. Diamagnetism is typically very low for common substances, requiring very strong magnetic fields to suspend even very small diamagnetic objects [21–23]. However, weak diamagnetic material may be used to stabilize the levitation of a permanent magnet [24]. The only strong diamagnetism is present in superconductors (SC) that force out a magnetic field (Meissner effect), and may be considered ideal diamagnetics with zero magnetic permeability. The absence of resistance to electric currents and perfect diamagnetism allow creating electromagnetic bearings with high stability parameters and the possibility to work with a turned off energy supply using non-decaying currents in superconducting coils, which makes superconductors especially interesting for use in non-contact bearings.

In order to provide levitation of a diamagnetic or superconductor, one has to compensate for gravity and inertia forces by a magnetic field force, defined by the field gradient product $B(\partial B/\partial z)$, where B is the induction of the magnetic field. Additionally, to provide mechanical stability, the field should possess a particular spatial configuration with an isolated minimum of potential energy, also referred to as a magnetic potential well (MPW) [25].

Field strength and its distribution in the space define the parameters of the devices utilizing the non-contact suspension, such as the mass of the levitated body, stiffness, overload capacity, torques, and so on. The proportional increase in the field strength increases the lifting force and stiffness. However, for a superconducting body there is a constraint: the magnetic field on the surface of the levitated body should always be less than the critical field H_{cr1}, and the current in the superconducting suspension coils, if they are used, must be less than the critical current, I_{cr}, otherwise the superconducting state breaks down [3]. Therefore, one can't just increase the field of support, but should find such spatial configuration of the sources that would provide the optimal characteristics of the suspension, while the constraints on the strength of the magnetic field and the currents in the superconducting coils are satisfied. Therefore, one needs a model that may be used to determine equilibrium, stability, and various characteristics of the superconducting suspensions for different spatial distributions of the sources, and to possibly allow for solving the problem of finding optimal source distributions [26].

Magnetic levitation systems are very interesting dynamical systems from both scientific and engineering points of view. Typically, the levitated body performs fast rotations and various rotational and translational oscillations are superimposed on this rotational motion. The dynamics of the levitated body in a non-contact suspension may be very complex due to the nonlinearities and coupling between translational and rotational motions, elasticity [27], and two-way electromechanical coupling with the support system. Therefore, one needs a model to study the various complex motions, energy transforms, and stability, which is complex enough to include these effects on the one hand. On the other hand, this model needs to be analytical or semi-analytical to allow a qualitative study of the dynamics. The existing analytical models are too simplistic. Typically, researchers consider dipole approximation for the levitated body with a quasi-homogeneous field, when the size of the levitated body is considered very small and it is replaced with a dipole with a magnetic dipole moment dependent on the external field [20,22,28], thus excluding from consideration all the effects related to the size and shape of the levitated body and field inhomogeneity, which are important for applications. In particular, the size of the body affects its stability in a magnetic field [29], the shape of the body affects its equilibrium, and a change of shape or rotations of a non-spherical body may cause parametric oscillations [30]. Even for close to spherical (quasi-spherical) levitated bodies, it is important to be able to account for the effects of non-sphericity and related torques, to be able to properly describe rotational motions, and estimate the precision of devices such as gyroscopes [31].

In this chapter, we present a model based on a calculation of the interaction energy between a spherical superconducting body of finite size and a supporting field created by a system of permanent magnets or electromagnets and represented in terms of expansions in spherical functions. The expressions for energy are obtained first in terms of expansions in spherical functions, in which case a finite number of terms in the sum may be calculated numerically. For some important cases (the case of N point magnetic charges and the case of N coaxial current loops), these expansions are converted into simple close-form analytical expressions. By placing N point charges or N coaxial current loops, one may construct rather complex source configurations, obtain various field distributions, and look for those configurations that provide the best characteristics for a particular device. Simple analytical expressions for energy allow the calculation of force characteristics, determine stability, study qualitatively the dynamics of the levitated body, see their dependence on the distributions of the field sources, and pose an inverse problem: find the source distribution providing the desired characteristics of the suspension. Even though the derivations are done for a spherical body, it is possible to consider small deviations from the spherical shape of the levitated body [30] and investigate the effects of non-sphericity on the dynamics of the levitated body [32–34]. The model may be useful to find distributions of sources providing a stable and robust equilibrium while satisfying mechanical and potentially thermal demands on the entire system. We use a quasi-static approximation, excluding the electromagnetic waves from the consideration [35].

6.2 Superconductor in Magnetic Field

Superconducting materials may be considered as perfect diamagnetics with zero permeability $\mu = 0$ due to the flux exclusion, known as the Meissner effect [3], causing repulsive forces exerted on superconductors placed in a magnetic field, pushing it to regions with a weaker magnetic field. The screening of magnetic flux is accompanied by persistent surface currents **i**, while inside the superconductor magnetic induction **B** is identically zero and no macroscopic currents may exist. The surface currents are not produced by an electric field, but arise by virtue of the discontinuity in the tangential component of the induction at the boundary of the superconducting body.

The tangential component of a magnetic field experiences discontinuity on the interface with the surface current [35]:

$$\mathbf{n} \times (\mathbf{H}_e - \mathbf{H}_i) = \mathbf{i}, \qquad (6.1)$$

where:

  $\mathbf{n}$ is the normal vector to the surface of the body
  $\mathbf{H}_e, \mathbf{H}_i$ is the magnetic field outside and inside the body, respectively

Since $\mathbf{B}_i = 0$ inside a superconductor, the density of the surface current is

$$\mathbf{i} = \mathbf{n} \times \mathbf{H}_e. \qquad (6.2)$$

The magnetic field near the current-carrying coils in the presence of superconductors can be calculated as the field in free space, created by bound surface currents $\mathbf{i}_k$ and the currents of the magnetizing coils $\mathbf{j}_k$ [35]:

$$\mathbf{H}(Q) = -\frac{\mu_0}{4\pi} \sum_k^L \int_{S_k} \left[\mathbf{i}_k(M_k) \times \nabla_Q \frac{1}{r_{QM_k}} \right] dS_{M_k} - \frac{\mu_0}{4\pi} \sum_k^N \int_{V_k} \left[\mathbf{j}_k(P_k) \times \nabla_Q \frac{1}{r_{QP_k}} \right] dV_{P_k}, \qquad (6.3)$$

The forces and torques acting on a superconductor in a magnetic field are due to the interaction of the induced surface currents with the external magnetic field. Thus, for known bound currents, $\mathbf{i}_k$ gives the formal solution of the problem. However, because the bound currents are not known in advance and are determined by the source currents, the problem reduces to determining the unknown superconducting sources (surface currents $\mathbf{i}$ or magnetic charges on the superconducting surfaces). This may be done by devising a system of Fredholm integral equations of the second kind in the densities of the secondary sources. By solving these integral equations and finding the distribution of the secondary sources, one can, by simple integration, calculate the entire field and then the force characteristics. For a spherical body, based on integral representation for vector potential [36], it is possible to find the Green function of the system made up of the sphere and the current-carrying coils, which solves integral equations by integrating the coil-type suspension with screens (or without them) placed close to the spherical rotor [37,38]. In other cases, numerical calculations are required; details of a numerical approach may be found in [39–45].

For the case without screens with constant currents in the sources, one may find the magnetic field by solving Laplace's equation for magnetic potential, using the perfect diamagnetism of a superconductor and satisfying the boundary conditions on its surface. We will use this approach, restricting ourselves to the case of suspension without screens.

6.3 Solving Field Problem

6.3.1 Secondary Field

Consider a homogeneous body S with relative magnetic permeability μ, placed in a known (in the absence of S) magnetic field $\mathbf{H}_0$, generated by a system of currents **j**. The total magnetic field **H** is the sum of the "primary" field **H**₀ and the "secondary" field **H**' due to the presence of body S. The very general case of a magnetized body with a continuously varying permeability magnetic field satisfies Maxwell equations [35]:

$$\nabla \times \mathbf{H} = \nabla \times (\mathbf{H}_0 + \mathbf{H}') = \mathbf{j}, \quad \nabla \cdot \mathbf{B} = 0, \quad (6.4)$$

whereas the primary field **H**₀ is

$$\nabla \times \mathbf{H}_0 = \mathbf{j}, \quad \nabla \cdot \mathbf{H}_0 = 0. \quad (6.5)$$

Subtracting the first Equation 6.5 from the first Equation 6.4, we have

$$\nabla \times \mathbf{H}' = 0, \quad (6.6)$$

that is, the secondary field has potential, and

$$\mathbf{H}' = -\nabla \varphi', \quad (6.7)$$

where φ' is the magnetic potential of the "secondary" field, and the problem of field determination in a suspension volume with a known external field reduces to finding the scalar function. This is true for an arbitrary-shaped body S.

Potential φ' satisfies Laplace's equation:

$$\Delta \varphi' = 0, \quad (6.8)$$

with boundary conditions following from the continuity conditions for the normal component of the flux density and the tangential component of the field on S:

$$\varphi'_i = \varphi'_e, \quad \mu \frac{\partial \varphi'_i}{\partial n} - \frac{\partial \varphi'_e}{\partial n} = (1-\mu)(\mathbf{H}_0 \cdot \mathbf{n}), \quad (6.9)$$

where:
- **n** is the vector of the outer normal to S
- φ'_i, φ'_e are the potentials of the secondary field inside and outside the body S

Furthermore, it is required that potential $\varphi' \to 0$ at infinity, and have no singularities in the origin. We assumed permeability of the outer space $\mu_{ext} = 1$.

6.3.2 Integral Equations for Secondary Field

After adding $(1-\mu)\partial \varphi'_i / \partial n$ to both parts of Equation 6.9, it may be rewritten in the form:

Force Characteristics of a Magnetic Suspension of an SC Sphere

$$\varphi'_i = \varphi'_e, \quad \frac{\partial \varphi'_i}{\partial n} - \frac{\partial \varphi'_e}{\partial n} = \sigma_M, \qquad (6.10)$$

where $\sigma_M = (1-\mu)[(\mathbf{H}_0 \cdot \mathbf{n}) + (\partial \varphi'_i / \partial n)]$

The Laplace equation with boundary conditions (Equation 6.10) defines a problem of determining the field inside and outside the body S with a charge distributed on its surface with density σ_μ. Thus, the potential may be formally expressed as

$$\varphi'_i = \frac{1}{4\pi} \int_S \frac{\sigma_\mu}{r_{QM}} ds = \frac{(1-\mu)}{4\pi} \int_S \frac{(\mathbf{H}_0 \mathbf{n})}{r_{QM}} ds + \frac{(1-\mu)}{4\pi} \int_S \frac{1}{r_{QM}} \frac{\partial \varphi'_i}{\partial n} ds, \qquad (6.11)$$

where r_{QM} is the distance between point Q inside the body, where the potential is calculated, and variable point M on the surface S. From the second Green's identity and the identity $\Delta(1/r_{QM}) = -4\pi\delta(Q-M)$, where Δ is Laplacian and δ is Dirac's delta-function [35], it follows that

$$\int_S \frac{1}{r_{QM}} \frac{\partial \varphi'_i}{\partial n} ds = 4\pi \varphi'_i(Q) + \int_S \varphi'_i \frac{\partial}{\partial n} \frac{1}{r_{QM}} ds. \qquad (6.12)$$

Substituting Equation 6.12 into Equation 6.11 results in an integral equation for the secondary field inside the body S:

$$\varphi'_i(Q) = \frac{(1-\mu)}{4\pi\mu} \int_S \frac{(\mathbf{H}_0 \mathbf{n})}{r_{QM}} ds + \frac{1-\mu}{4\pi\mu} \int_S \varphi'_i(M) \frac{\partial}{\partial n}\left(\frac{1}{r_{QM}}\right) ds. \qquad (6.13)$$

Similarly, the integral equation for the "secondary" field outside the body S will be

$$\varphi'_e(Q) = \frac{(1-\mu)}{4\pi} \int_S \frac{(\mathbf{H}_0 \mathbf{n})}{r_{QM}} ds + \frac{(1-\mu)}{4\pi} \int_S \varphi'_e(M) \frac{\partial}{\partial n}\left(\frac{1}{r_{QM}}\right) ds. \qquad (6.14)$$

Here, the point Q is located outside the body S.

6.3.3 Solution for Spherical Shape

Equations 6.13 and 6.14 are derived for the arbitrary-shaped body S. Now assume that S has a spherical shape and put the origin in the center of S, then expand r_{QM}^{-1} in the series of spherical functions [35]:

$$\frac{1}{r_{QM}} = 4\pi \sum_{l,m} \frac{r^l}{(2l+1)r'^{l+1}} Y_{lm}(\theta, \varphi) Y^*_{lm}(\theta', \varphi'), \qquad (6.15)$$

where primed and unprimed angles stay for points M and Q, respectively. Substituting Equation 6.15 into Equation 6.13, we obtain

$$\varphi'_i = \frac{1-\mu}{\mu}\sum_{l,m}\frac{r^l Y_{lm}(\Omega)}{(2l+1)R^{l-1}}\int_S \mu_0(\mathbf{H}_0\mathbf{n})Y^*_{l,m}(\Omega')ds - \frac{1-\mu}{\mu}\sum_{l,m}\frac{(l+1)}{2l+1}\frac{r^l Y_{lm}(\Omega)}{R^l}\int_S \varphi'_i Y^*_{l,m}(\Omega')ds$$

$$= \frac{1-\mu}{\mu}\sum_{l,m}\frac{r^l Y_{l,m}(\Omega)}{(2l+1)R^{l+1}}E^i_{lm} - \frac{1-\mu}{\mu}\sum_{l,m}\frac{(l+1)}{2l+1}\frac{r^l}{R^{l+2}}Y_{l,m}(\Omega)C^i_{lm}, \quad (6.16)$$

where

$$E^i_{lm} = \int_S (\mathbf{H}_0\mathbf{n})Y^*_{l,m}(\Omega^*)ds, \quad (6.17)$$

$$C^i_{lm} = \int_S \varphi'_i Y^*_{lm}(\Omega')ds, \quad (6.18)$$

R is the radius of the sphere and Ω is the set of angles θ, φ.

Substituting Equation 6.16 into Equation 6.18 and integrating over a unit sphere gives

$$C^i_{lm} = \frac{E^i_{lm}R}{2l+1} - \frac{1-\mu}{\mu}\frac{l+1}{2l+1}C^i_{lm},$$

therefore,

$$C^i_{lm} = \frac{(1-\mu)E^i_{lm}R}{\mu l + l + 1}. \quad (6.19)$$

Then, the potential could be written in the following form:

$$\varphi'_i = \int_S (1-\mu)(\mathbf{H}_0\mathbf{n})\Phi^i(Q,M)ds, \quad (6.20)$$

where

$$\Phi^i(Q,M) = \sum_{l,m}\frac{1}{l\mu+l+1}\frac{r^l}{R^{l+1}}Y_{lm}(\Omega_Q)Y^*_{lm}(\Omega_M) \quad (6.21)$$

is the potential inside the sphere with magnetic permeability μ due to a unit point charge on its surface.

For the external potential of the secondary field, the formula is

$$\varphi'_e(Q) = \int_S (1-\mu)(\mathbf{H}_0\mathbf{n})\Phi^e(Q,M)ds, \qquad (6.22)$$

where $\Phi^e(Q,M)$ is obtained from $\Phi^i(Q,M)$ by switching R and r. In the simplest case of a uniform external field, Equations 6.20 and 6.21 yield the well-known result [46]:

$$\varphi'_i = \frac{1-\mu}{\mu+2}(\mathbf{H}r), \quad \varphi'_e = \frac{1-\mu}{\mu+2}\frac{R^3}{r^3}(\mathbf{H}_0 r). \qquad (6.23)$$

6.3.4 Total Field

It is convenient to represent the potential of the primary field $\mathbf{H}_0 = -\nabla\varphi_0$ in terms of spherical functions Y_{lm}:

$$\varphi_0 = \sum_{l,m}\sqrt{\frac{4\pi}{2l+1}}a_{lm}r^l Y_{lm}(\Omega). \qquad (6.24)$$

Such an expansion is always possible since the potential of the primary field in any point outside the field sources, where $\nabla\times\mathbf{H}_0 = 0$, satisfies the Laplace equation. The expansion coefficients a_{lm} could be calculated analytically or numerically for a known sources distribution; they depend on the sources' geometry, the distances between sources and origin, and the currents' values. If the field has an axis of symmetry that coincides with the polar axis of a spherical system of reference, then $a_{lm} = 0$ for all $m \neq 0$.

Substituting Equation 6.24 into Equation 6.17 and integrating gives

$$E^i_{lm} = \sqrt{\frac{4\pi}{2l+1}}a_{lm}lR^{l+1}, \qquad (6.25)$$

it is obvious that $E^l_{lm} = E^i_{lm}$. Therefore, the potential of the total field inside the sphere is

$$\varphi^i = \varphi_0 + \varphi'_i = \sum_{l,m}\sqrt{\frac{4\pi}{2l+1}}\frac{2l+1}{l+1+l\mu}a_{lm}r^l Y_{lm}(\Omega), \qquad (6.26)$$

and outside the sphere is

$$\varphi^e = \varphi_0 + \varphi'_e = \sum_{l,m}\sqrt{\frac{4\pi}{2l+1}}a_{lm}\left(r^l + \frac{(1-\mu)l}{l\mu+l+1}\frac{R^{2l+1}}{r^{l+1}}\right)Y_{lm}(\Omega). \qquad (6.27)$$

The case of a superconductor follows from these equations when $\mu = 0$ and only the potential outside the sphere has a physical sense; for ferromagnetic $\mu = \infty$.

6.4 Energy, Force Characteristics, and Dynamics

6.4.1 Energy Integrals

When the total magnetic field **H** and currents **i** on the surface of a superconductor are known, one may use them to determine the forces and torques exerted on the superconductor and then plug the forces and torques into the equations of motion. However, instead of dealing with vector quantities of force and torque, it is convenient to calculate the first scalar quantity, the potential energy of interaction, and then use it to define the force characteristics and use Lagrange formalism to study dynamics.

When surface currents on a levitated superconducting body and currents in the support system are known, one may use the integral [35]:

$$W = \frac{1}{2}\int_V \mathbf{A} \cdot \mathbf{j} dV = \sum_{i=1}^N \frac{\mu_0}{8\pi} \int_{V_i} dV_1 \sum_{j=1}^N \int_{V_2} dV \frac{(\mathbf{j}_i \cdot \mathbf{j}_j)}{|\mathbf{x}_i - \mathbf{x}_j|_2}. \tag{6.28}$$

It may also be expressed in terms of total currents I and their integral geometrical characteristics inductances:

$$W = \frac{1}{2}\sum_{j,k} L_{jk} I_j I_k, \quad L_{ij} = \frac{\mu_0}{4\pi I_i I_j} \int_{C_i} dV_i \int_{C_j} dV_j \frac{\mathbf{j}_i \mathbf{j}_j}{|\mathbf{x}_i - \mathbf{x}_j|}. \tag{6.29}$$

Alternatively, one may use the expression for a magnetized body in an external field (the case of linear isotropic media) [35]:

$$W = -\frac{1}{2}\int_V \mathbf{M} \cdot \mathbf{B}_0 dV = \frac{1}{2}\int_V \mu_0(1-\mu)\mathbf{H} \cdot \mathbf{H}_0 dV = \frac{1}{2}\int_V \mu_0(1-\mu)\nabla\varphi^i \nabla\varphi_0 dV, \tag{6.30}$$

where:
  B = $\mu_0\mu\mathbf{H}$ is magnetic induction
  M = $(\mu - 1)\mathbf{H}$ is the magnetization per unit volume
  μ = constant, in terms of potentials

$$W = \frac{1}{2}\int_V \mu_0(1-\mu)\nabla\varphi^i \nabla\varphi_0 dV. \tag{6.31}$$

Using Green's identity

$$\int_V \nabla \varphi^i \nabla \varphi_0 dV = \int_S \varphi^i \frac{\partial \varphi_0}{\partial n} dS - \int_V \varphi^i \Delta \varphi_0 dV$$

and condition $\Delta \varphi_0 = 0$ inside the levitated body, energy (Equation 6.30) may be represented as

$$W = \frac{\mu_0(1-\mu)}{2} \int_S \varphi^i \frac{\partial \varphi_0}{\partial n} dS. \tag{6.32}$$

Substitution Equations 6.24 and 6.27 into Equation 6.32 and integrating over the sphere, using the orthogonality of the spherical functions, results in the expression for the interaction energy in terms of the expansion coefficients of the primary field [29]:

$$W = 2\pi\mu_0(1-\mu) \sum_{l,m} \frac{lR^{2l+1}}{l+1+l\mu} a_{lm} a_{lm}^*. \tag{6.33}$$

In [47], the interaction energy is calculated in terms of the integral from the field of suspension in the absence of the sphere:

$$W = \mu_0 \alpha \int_V \left(\frac{r'}{R}\right)^\alpha H_0^2(|\boldsymbol{\rho}+\boldsymbol{\rho}'|) dV, \tag{6.34}$$

where:
$\alpha = 1-\mu/1+\mu$
$\boldsymbol{\rho}$ is the vector from the center of mass of the sphere to some fixed point in space, in particular it can be the center of suspension

It can be shown that Equations 6.33 and 6.34 are equivalent. Expression 6.33 is more convenient for calculating energy in terms of expansion coefficients, while Equation 6.34 is convenient for deriving forces and stiffness.

For the case when the levitated body is very small, one may neglect the change of the field over the extent of the levitated body (quasi-homogeneous field) and take it out of the integral in Equation 6.34, in which case:

$$W = \mu_0 \alpha H_0^2(|\boldsymbol{\rho}+\mathbf{r}'|) \int_V \left(\frac{r'}{R}\right)^\alpha dV = \frac{4\pi\mu_0\alpha}{3+\alpha} R^3 H^2(\boldsymbol{\rho}). \tag{6.35}$$

This expression coincides with the expression used in [20] and for a weak diamagnetic, that is, small $\chi = \mu - 1$, reduces to the approximate expressions used by [22,28], $W = -\chi B_0^2 V/2\mu_0$.

6.4.2 Forces

Forces may be calculated directly from Equation 6.33 by differentiation, in which case the derivatives of the coefficient a_{lm} must be calculated. It is, however, possible to calculate the derivatives of **H** in terms of coefficient a_{lm} using the Clebsch–Gordan coefficients formalism. We will use Expression 6.34 to calculate the forces between a magnetized sphere and a support field. The force components in a cyclic coordinates system [48] are as follows:

$$F_\nu = -\frac{\partial}{\partial x_\nu} W(\rho) = -\mu_0 \alpha \frac{\partial}{\partial x_\nu} \int_V \left(\frac{r'}{R}\right)^\alpha H_0^2(|\rho + \mathbf{r}'|) dV \quad (6.36)$$

$$= -\mu_0 \alpha \int_V \left(\frac{r'}{R}\right)^\alpha \frac{\partial}{\partial x_\nu} H_0^2(|\rho + \mathbf{r}'|) dV = -\mu_0 \alpha \int_V \left(\frac{r'}{R}\right)^\alpha \frac{\partial'}{\partial x_\nu} H_0^2(|\rho + \mathbf{r}'|) dV.$$

The derivative without prime corresponds to differentiation with respect to ρ, while the derivative with prime corresponds to differentiation with respect to r'. Index ν runs through –1, 0, 1, and

$$F_0 = F_z, \quad F_1 = -\frac{1}{\sqrt{2}}(F_x + iF_y), \quad F_{-1} = \frac{1}{\sqrt{2}}(F_x - iF_y). \quad (6.37)$$

Calculating the derivative of the square of the magnetic field $(\partial'/\partial x_\nu)H_0^2$ and substituting it into Equation 6.36 yields [35]

$$F_\nu = 4\pi\mu_0 \sum_{l,m} \frac{(\mu-1)lR^{2l+1}}{l+1+l\mu} \sqrt{\frac{(l+1+m+\nu)!(l+1-m-\nu)}{(1-\nu)!(1+\nu)!(l-m)!(l+m)!}} a_{lm} a^*_{l+1,m+\mu}. \quad (6.38)$$

From Equation 6.38, it follows that force differs from zero only if field harmonics with neighboring l are not zeros. In particular, if the field of suspension has a symmetry or antisymmetry axis, that is, there are only even or uneven harmonics in the expansion, force vanishes.

For the case of superconductor $\mu = 0$, we have

$$F_\nu = -4\pi\mu_0 \sum_{l,m} \frac{l}{l+1} \sqrt{\frac{(l+1+m+\nu)!(l+1-m-\nu)}{(1-\nu)!(1+\nu)!(l-m)!(l+m)!}} R^{2l+1} a_{lm} a^*_{l+1,m+\mu} \quad (6.39)$$

It is clear from physical considerations that in the case of an axisymmetric field when the center of the sphere is on the symmetry axis, then the only possible non-vanishing component is F_z. In this case, all coefficients with a

Force Characteristics of a Magnetic Suspension of an SC Sphere 177

non-zero second index vanish in the potential expansion and Equation 6.38 reduces to

$$F_z = 4\pi\mu_0 (\mu - 1) \sum_{l=1}^{\infty} \frac{l(l+1)}{l+1+l\mu} R^{2l+1} a_l a_{l+1}. \quad (6.40)$$

6.4.3 Stiffness

Using Equation 6.35, one may calculate stiffness as follows:

$$k_{\nu\lambda} = -\frac{\partial F_\nu}{\partial x_\lambda} = \alpha\mu_0 \int_V \left(\frac{r'}{R}\right) \frac{\partial^2}{\partial x_\lambda \partial x_\nu} H_0^2 dV, \quad (6.41)$$

where ν and λ run through values 0, 1, –1. Due to the conservative nature of forces, the stiffness tensor $k_{\nu\lambda}$ is symmetrical:

$$k_{0,0} = k_{zz}$$

$$k_{0,1} = k_{10} = -\frac{1}{\sqrt{2}}(k_{zx} + ik_{zy})$$

$$k_{0,-1} = k_{-1,0} = -k_{01}^* \quad (6.42)$$

$$k_{1,-1} = k_{-1,1} = -\frac{1}{\sqrt{2}}(k_{xx} + k_{yy})$$

$$k_{-1,-1} = k_{1,1}^*$$

and

$$k_{xx} = \frac{1}{2}(k_{1,1} + k_{-1,-1} - 2k_{1,-1}), \quad k_{yy} = -\frac{1}{2}(k_{1,1} + k_{-1,-1} + 2k_{1,-1}). \quad (6.43)$$

For an axisymmetric field, the expressions for stiffness simplify, since the field is described by the expansion, where the coefficients with a second non-zero index vanish. In this case, expressions for stiffness reduce to [49]

$$k_{zz} = 4\pi\mu_0 (1-\mu) \sum_{l=1}^{\infty} \frac{l(l+1)}{l+1+l\mu} R^{2l+1} \{(l+1)a_{l+1}^2 + (l+2)a_l a_{l+2}\}, \quad (6.44)$$

$$k_{xx} = k_{yy} = k_\perp = 4\pi\mu_0 \frac{1-\mu}{2} \sum_l \frac{l(l+1)}{l+1+l\mu} R^{2l+1} \{la_{l+1}^2 - (l+2)a_l a_{l+2}\}. \quad (6.45)$$

For stable equilibrium, components k_{zz}, k_{xx}, and k_{yy} must be positive. A summation stiffness components of k_{xx}, k_{yy}, and k_{zz} yields

$$k_{zz} + k_{xx} + k_{yy} = 4\pi\mu_0(1-\mu)\sum_{l,m}\frac{lR^{2l+1}}{l+1+l\mu}(l+1)(2l+1)|a_{l+1,m}|^2 \qquad (6.46)$$

In the case of a diamagnetic or a superconductor ($\mu < 0$), the sum of the stiffness components is positive, and it may be possible that each component is positive and that conservative stability is possible in the suspension, which is consistent with the Braunbeck–Earnshaw theorem [19].

6.4.4 Dynamics

With energy, forces, and torques calculated, one may study the dynamics of the levitated body in the suspension. It is convenient to use the Lagrange–Maxwell formalism [3,50], which states that the motion in an electromechanical system may be deduced from the Lagrange function. The Lagrange function for electromechanical system is

$$L = T(q_i,\dot{q}_i) - \Pi(q_i) + W_m(q_i,i_k), \qquad (6.47)$$

where:
- $q_i, \dot{q}_i$ are the generalized coordinates and velocities of the mechanical part
- i_k are the source currents
- T and Π are the kinetic and potential energy
- W_m is the magnetic and electrostatic interaction energy

Kinetic energy T of a rigid body consists of translational and rotational parts [51]:

$$T = \frac{mv^2}{2} + \frac{1}{2}I_{ik}\Omega_i\Omega_k, \qquad (6.48)$$

where:
- Ω_i are the components of angular velocity
- I_{ik} are the components of inertia tensor

The potential energy assuming a vertical z-axis will be

$$\Pi = mgz. \qquad (6.49)$$

For constant current sources, the magnetic energy must be related to the kinetic energy, and for constant fluxes it is related to the potential energy. Then, the equation of motion of the rotor in the superconducting suspension will have the form of a Lagrange equation:

$$\frac{d}{dt}\frac{\partial L}{\partial q_S} - \frac{\partial L}{\partial q_S} = Q_{qs}, \qquad (6.50)$$

where Q_{qS} is the generalized force of a non-conservative nature, conjugated to the generalized coordinate q_S. The case of constant flux through the coil implies the presence of cyclic integrals in the system; therefore, in addition to the Lagrange–Maxwell function, we can construct the Routh–Maxwell function. Clearly, the two means of constructing the equation are equivalent.

6.5 Suspension on Circular Current Loops

6.5.1 System of Circular Current Loops

A typically supporting field is created by electromagnets or superconducting circuits [3]. Assume that magnetic field **H** is created by a system of coaxial current loops with radius r_i, carrying currents $\varepsilon_i I_i$, where $\varepsilon_i = \pm 1$ and I_i is the current magnitude, $i = 1, 2, \ldots$. Then, a spherical coordinate system with its origin in the center of the sphere magnetic field potential may be written as

$$\varphi_0 = \sum A_l r^l P_l(\cos\theta), \qquad (6.51)$$

where

$$A_l = -\sqrt{\pi\mu_0} \sum \varepsilon_i \frac{I_i(\sin\alpha_i)^{l+1}}{r_i^l l} P_l^1(\cos\alpha_i) \qquad (6.52)$$

and $2\alpha_i$ is an angle in which the diameter of the ith loop is seen from the origin of the coordinate system. Substituting Equation 6.52 into Equation 6.33 and taking $\mu = 0$, results in the following expression of energy:

$$W = \frac{\pi R}{2}\mu_0 \sum_{i,k}\frac{I_i I_k r_i r_k}{b_i b_k}\sum_{l=1}^{\infty}\frac{P_l^1(\cos\alpha_i)P_l^1(\cos\alpha_k)}{l(l+1)}. \qquad (6.53)$$

Summing Equation 6.53, yields [49]

$$W = \frac{\mu_0}{2}\sum_{i,k} I_i I_k \sqrt{r_i r_k}\, k_{i,k}^3 C(k_{ik}), \qquad (6.54)$$

where

$$k_{ik}^2 = \frac{4R^2 r_i r_k}{R^4 - 2R^2(z_i z_k - r_i r_k) + b_i^2 b_k^2}, \qquad (6.55)$$

$$C(k_{ik}) = \frac{1}{k_{ik}^4}\left[(2 - k_{ik}^2)K(k_{ik}) - 2E(k_{ik})\right]. \qquad (6.56)$$

K, E are elliptic integrals.

6.5.2 One Current Loop

Consider a superconducting sphere in the field of a single circular current loop of radius r_b, as shown on Figure 6.1. The energy of interaction with one current loop is obtained from Equation 6.54 when $i = k = 1$:

$$W = \frac{\mu_0}{2} I^2 r_b k^3 C(k), \qquad (6.57)$$

where

$$k^2 = \frac{4R^2 r_b^2}{\left(R^2 + r_b^2\right)^2 + 2\left(r_b^2 - R^2\right)z^2 + z^4}. \qquad (6.58)$$

From Equation 6.57 it is now easy to find the force acting on the sphere and the longitudinal stiffness.

The transversal stiffness is much harder to calculate, as it is not possible to obtain a simple analytical expression even for one current loop and the sum (Equation 6.45) must be calculated numerically. In the case of a small sphere, the sum converges very fast and it is sufficient to take only few terms. The

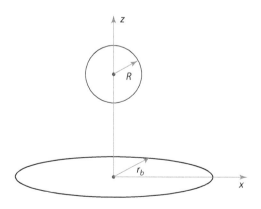

FIGURE 6.1
A superconducting sphere in the field of a circular loop.

Force Characteristics of a Magnetic Suspension of an SC Sphere

dependence of stiffness on the parameter $\eta = z/r_b$, characterizing the relative displacement of the sphere from the loops plane, is depicted in Figure 6.2.

The analytical dependence of force and stiffness when $R/b \ll 1$ has the form:

$$F^C = -F_0^C \frac{\eta}{(1+\eta^2)^4}, \quad k_z^C = \frac{F_0^C}{r_b} \frac{7\eta^2 - 1}{(1+\eta^2)^5}, \quad k_\perp^C = \frac{F_0^C}{4r_b} \frac{2 - 5\eta^2}{(1+\eta^2)^5}, \quad (6.59)$$

where

$$F_0^C = \frac{3\pi\mu_0 I^2}{2}\left(\frac{R}{r_b}\right)^3, \quad \eta = \frac{z}{r_b}.$$

From Equation 6.59, it follows that in the interval:

$$0.378 < \eta < 0.63$$

all stiffness components are positive and suspension is conservatively stable. The weight of the body will be restricted by the limits:

$$0.97 < \frac{mg}{F_0} < 1.12$$

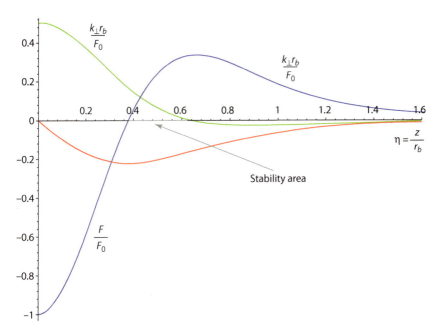

FIGURE 6.2
Dependence of energy, force, and stiffness on the parameter η.

so that the equilibrium point fits the stability interval. If it is too heavy, then it will fall through the ring, if it is too light, then it will lift up and slide sideways, since transverse stiffness becomes negative above the point $\eta = 0.63$.

With the increase in the size of the sphere, the stability regions change, for example, for $R = 0.56 r_b$ the stability region is

$$0.435 < \eta < 0.767.$$

6.6 Suspensions on Permanent Magnets

6.6.1 Permanent Magnets and Magnetic Charges

We have previously discussed suspensions created by a system of current loops. However, a supporting magnetic field may also be generated by a system of permanent magnets. An important advantage of such a setup is that the support is truly passive as no energy is required to feed electromagnets. Moreover, such a system is very simple and reliable.

Recent technological advances have made it possible to create rare earth elements (REE)-based permanent magnets generating a field that is sufficient for levitation. Appropriately processed, REE compounds, such as $SmCo_5$ and $NdCo_5$, provide coercive forces as high as 1.4 T (the highest known value for this type of material) and a giant magnetic energy of up to 10^4 T A/m, which is two orders of magnitude higher than the corresponding energy for magnets made of iron-group metals. This makes it possible to produce magnets several tens of times stronger than those based on iron-group metals [52].

Permanent magnets are characterized by remnant magnetization B_r, which reaches 1.4 T in modern magnets [52,53].

A magnetic field created by permanent magnets may be effectively modeled using the concept of a magnetic charge, even though magnetic charges do not exist in nature. In the following, we consider a superconducting sphere in the magnetic field generated by a system of permanent cylindrical magnets with axial magnetization. One pole of each of the magnets faces the suspended body. The field produced by a thin cylindrical magnet (magnetic needle, Figure 6.3) may be viewed as a dipole field [25,54]. For sufficiently

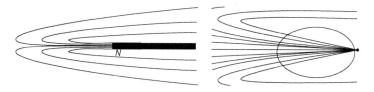

FIGURE 6.3
Thin magnet and its field near one end.

long magnets, the field of one pole can be assumed to be negligible near the other pole (the encircled region in Figure 6.3). Therefore, we assume that the field in this region is produced by N point magnetic poles.

A charge corresponding to one of the poles of a cylindrical magnet can be estimated by the formula $q = \pi a^2 M$ [25,54], where a is the radius of the cylinder and M is the magnetization of the magnet. The magnetization of a cylindrical magnet is related to the remnant induction by the expression $M = B_r / \mu_0$ [35,46,54], thus $q = \pi a^2 B_r / \mu_0$.

6.6.2 Force Function of a Sphere in a Suspending Field Produced by N Point Magnetic Poles

At some point B outside the sphere S, potential φ_0 of a set of N magnetic poles with spherical coordinates b_i, θ_i, and φ_i ($i = 1, 2, \ldots, N$) originating at the center of the sphere can be written in the following form [35]:

$$\varphi_0 = \frac{\mu_0}{4\pi} \sum_{i=1}^{N} \frac{q_i}{r_{BA_i}} = \frac{\mu_0}{4\pi} \sum_{i=1}^{N} \frac{q_i}{\sqrt{r^2 + b_i^2 - 2rb_i(\cos\theta\cos\theta_i + \sin\theta\sin\theta_i\cos(\varphi - \varphi_i))}}$$

$$= \sum_{i=1}^{N} \sum_{l,m} \frac{\mu_0}{(2l+1)} r^l \frac{q_i}{b_i^{l+1}} Y_{lm}(\theta, \varphi) Y^*(\theta_i, \varphi_i)$$

$$= \sum_{l,m} \sqrt{\frac{1}{2l+1}} r^l Y_{lm}(\theta, \varphi) \left\{ \sum_{i=1}^{N} \sqrt{\frac{1}{2l+1}} \frac{\mu_0 q_i}{b_i^{l+1}} Y^*(\theta_i, \varphi_i) \right\} \quad (6.60)$$

$$= \sum_{l,m} a_{l,m}^* \sqrt{\frac{1}{(2l+1)}} r^l Y_{lm}(\theta, \varphi),$$

where

$$a_{l,m}^* = \sum_{i=1}^{N} \sqrt{\frac{1}{(2l+1)}} \frac{\mu_0 q_i}{b_i^{l+1}} Y^*(\theta_i, \varphi_i). \quad (6.61)$$

Here, r, θ, and φ are the spherical coordinates of point B.

Due to the principle of superposition, one can create a rather complex field from N magnetic poles. For example, from Equations 6.60 and 6.61 it is easy to obtain the potential of the field, generated by a magnetic ring, which can be considered as consisting of an infinite number of poles, or a set of magnetic rings. For this purpose, one has to integrate Equation 6.61 over φ from 0 to 2π, assuming that the poles are distributed evenly along the rings. Then, Equation 6.61 reduces to

$$a_{l0} = \frac{\mu_0}{\sqrt{4\pi}} \sum_{i=1}^{N} \frac{q_i}{b_i^{l+1}} P_l(\cos\theta_i), \tag{6.62}$$

where $P_l(\cos\theta)$ are the Legendre polynomials.

Substituting Equation 6.61 into Equation 6.33 yields

$$\begin{aligned} W &= \frac{\mu_0(1-\mu)}{2} \sum_{l,m} \frac{lR^{2l+1}}{l+1+l\mu} \sum_{i,k} \frac{1}{2l+1} \frac{q_i q_k}{(b_i b_k)^{l+1}} Y_{l,m}^*(\theta_i,\varphi_i) Y_{l,m}(\theta_k,\varphi_k) \\ &= \frac{\mu_0(1-\mu)}{8\pi} \sum_{i,k} \sum_{l} \frac{lR^{2l+1}}{l+1+l\mu} \frac{q_i q_k}{(b_i b_k)^{l+1}} P_l(\cos\theta_{ik}), \end{aligned} \tag{6.63}$$

where

$$\cos\theta_{ik} = \cos\theta_i \cos\theta_k + \sin\theta_i \sin\theta_k \cos(\varphi_i - \varphi_k). \tag{6.64}$$

This expression holds for arbitrary material: diamagnetic, paramagnetic, and ferromagnetic. The case of a superconductor corresponds to $\mu = 0$, the case of a ferromagnetic corresponds to $\mu = \infty$. For arbitrary μ, the summation in index l may not be converted to a closed-form expression. However, for $\mu = 0$ and $\mu = \infty$, it can be done.

Taking $\mu = 0$ for a superconducting rotor:

$$W = \frac{\mu_0}{8\pi} \sum_{i,k} q_i q_k \frac{R}{b_i b_k} \sum_{l=1}^{\infty} \frac{l}{l+1} \left(\frac{R^2}{b_i b_k}\right)^l P_l(\cos\theta_{ik}). \tag{6.65}$$

The sum may be converted to the closed-form expression [29]:

$$\begin{aligned} W &= \frac{\mu_0}{8\pi} \sum_{i=1}^{N} q_i^2 R \left\{ \frac{1}{b_i^2 - R^2} - \frac{1}{R^2} \ln \frac{b_i^2}{b_i^2 - R^2} \right\} \\ &+ \frac{\mu_0}{8\pi} \sum_{i<k} q_i q_k R \left\{ \frac{1}{\sqrt{(b_i^2 - R^2)(b_k^2 - R^2) + \eta_{ik}^2 R^2}} \right. \\ &\left. - \frac{1}{R^2} \ln \frac{2\sqrt{(b_i^2 - R^2)(b_k^2 - R^2) + \eta_{ik}^2 R^2} - (b_i^2 + b_k^2) + 2R^2 + \eta_{ik}^2}{\eta_{ik}^2 - (b_i - b_k)^2} \right\}. \end{aligned} \tag{6.66}$$

In this expression, $\eta_{ik}^2 = b_i^2 + b_k^2 - 2b_i b_k \cos\theta_{ik}$ is the distance between the ith and kth poles.

6.6.3 Two Point Magnets

Consider a superconducting sphere of radius R in the field of two magnetic poles with charge q and coordinates $z_1 = d$ and $z_2 = -d$, as shown in Figure 6.4.

For displacements along the y-axis, the potential energy:

$$W_\perp = \frac{\mu_0}{4\pi} \left(q^2 R \left\{ \frac{1}{d^2 + y^2 - R^2} - \frac{1}{R^2} \ln \frac{d^2 + y^2}{d^2 + y^2 - R^2} \right\} \right.$$

$$+ q^2 R \left\{ \frac{1}{\sqrt{(d^2 + y^2 - R^2)^2 + 4d^2 R^2}} \right.$$

$$\left. \left. - \frac{1}{R^2} \ln \frac{\sqrt{(d^2 + y^2 - R^2)^2 + 4d^2 R^2} + R^2 + d^2 - y^2}{2d^2} \right\} \right), \quad (6.67)$$

or, in the dimensionless form:

$$w_\perp = \frac{W_\perp}{W_0} = \left\{ \frac{\gamma}{1 - \gamma^2 + \varepsilon^2} - \frac{1}{\gamma} \ln \frac{1 + \varepsilon^2}{1 - \gamma^2 + \varepsilon^2} + \frac{\gamma}{\sqrt{(1 - \gamma^2 + \varepsilon^2)^2 + 4\gamma^2}} \right.$$

$$\left. - \frac{1}{\gamma} \ln \frac{\sqrt{(1 - \gamma^2 + \varepsilon^2)^2 + 4\gamma^2} + 1 + \gamma^2 - \varepsilon^2}{2} \right\}, \quad (6.68)$$

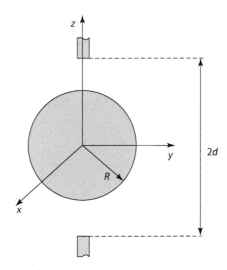

FIGURE 6.4
Superconductor of a spherical shape in the field of two magnetic poles.

where $\gamma = R/d$, $\varepsilon = y/d$, $W_0 = (\mu_0/4\pi)(q^2/d)$. Then, the associated force is given by

$$f_\perp = \frac{F_\perp}{F_0} = -\frac{\partial w}{\partial \varepsilon} = \frac{2\gamma\varepsilon}{(1-\gamma^2+\varepsilon^2)} + \frac{\left(\dfrac{2\varepsilon}{1-\gamma^2+\varepsilon^2} - \dfrac{2(1+\varepsilon^2)\varepsilon}{(1-\gamma^2+\varepsilon^2)^2}\right)(1-\gamma^2+\varepsilon^2)}{\gamma(1+\varepsilon^2)}$$

$$+ \frac{\gamma(4\varepsilon - 4\gamma^2\varepsilon + 4\varepsilon^3)}{2(1+2\gamma^2+2\varepsilon^2+\gamma^4-2\gamma^2\varepsilon^2+\varepsilon^4)^{3/2}} \qquad (6.69)$$

$$+ \frac{\dfrac{4\varepsilon-4\gamma^2\varepsilon+4\varepsilon^3}{4\sqrt{1+2\gamma^2+2\varepsilon^2+\gamma^4-2\gamma^2\varepsilon^2+\varepsilon^4}} - \varepsilon}{\gamma\left(\dfrac{\sqrt{1+2\gamma^2+2\varepsilon^2+\gamma^4-2\gamma^2\varepsilon^2+\varepsilon^4}}{2} + \dfrac{1}{2} + \dfrac{\gamma^2}{2} - \dfrac{\varepsilon^2}{2}\right)},$$

where $F_0 = \mu_0 q^2/(4\pi d^2)$, and stiffness

$$k_\perp = \frac{K_\perp}{K_0} = \frac{\partial^2 w_\perp}{\partial \varepsilon^2}, \quad K_0 = \frac{\mu_0}{4\pi}\frac{q^2}{d^3}. \qquad (6.70)$$

Expression 6.70 for arbitrary may be obtained in close form in Maple, but it is cumbersome. The first term of the expansion in powers of ε is

$$k_\perp = \frac{2\gamma^3(1-7\gamma^2+\gamma^4+\gamma^6)}{(\gamma^2-1)^2(\gamma^2+1)^3}. \qquad (6.71)$$

Expressions 6.70 and 6.71 allow one to estimate stiffness $k_\perp$ as a function of the parameter γ. This dependence is shown in Figure 6.5. Stiffness $k_\perp$ is positive for $\gamma < 0.37$ and then becomes negative.

Figure 6.6 plots the transverse stiffness against the displacement along the y-axis for $\gamma = 0.5$ and $\gamma = 0.32$ ($q = 3.5$ A m, $R = 0.001$ m, $d = 0.03$ m). The value of the charge, $q = 3.5$ A m, was obtained for $B_r = 1.4$ T and $a = 0.001$ m. Here, a is small compared to the rotor radius and the rotor–magnet distance; accordingly, the charge can be considered as a point charge.

The parameters of a suspension should be selected so that the rotor remains in the stable state under the action of inertial forces. Depending

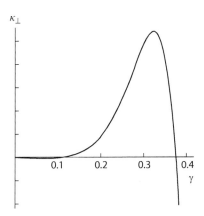

FIGURE 6.5
Dependence lateral stiffness on the parameter γ.

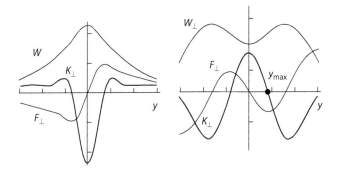

FIGURE 6.6
Energy, transverse force, and stiffness for the cases $\gamma = 0.5$ and $\gamma = 0.32$.

on a particular application, one may want to maximize or minimize (as is required for accelerometers) stiffness. From Equation 6.71 and Figure 6.6, it follows that the stiffness achieves maximum at $\gamma = 0.32$.

Stiffness also depends on the radius of the rotor. Expressing d in Equations 6.70 and 6.71 in terms of R:

$$k_\perp = \frac{\mu_0}{4\pi} \frac{q^2}{R^3} 2\gamma^6 \left(\frac{2}{(1+\gamma^2)^3} + \frac{1}{(1-\gamma^2)^2} \right) \qquad (6.72)$$

one may see that $k_\perp$ is inversely proportional to the cube of a radius. Note that this dependence may not be obtained in simpler models with a quasi-homogeneous approximation.

For longitudinal displacement, the potential energy is given by

$$W_{||} = \frac{\mu_0}{4\pi}\left(\frac{1}{2}q^2R\left\{\frac{1}{(d-z)^2-R^2} - \frac{1}{R^2}\ln\frac{(d-z)^2}{(d-z)^2-R^2}\right\}\right.$$

$$+ \frac{1}{2}q^2R\left\{\frac{1}{(d+z)^2-R^2} - \frac{1}{R^2}\ln\frac{(d+z)^2}{(d+z)^2-R^2}\right\}$$

$$\left.+q^2R\left\{\frac{1}{d^2-z^2+R^2} - \frac{1}{R^2}\ln\frac{d^2-z^2+R^2}{d^2-z^2}\right\}\right),$$

(6.73)

or, in the dimensionless form,

$$w_{||} = \frac{W_{||}}{W_0} = \frac{1}{2}\gamma\left\{\frac{1}{(1-\zeta)^2-\gamma^2} - \frac{1}{\gamma^2}\ln\frac{(1-\zeta)^2}{(1-\zeta)^2-\gamma^2} + \frac{1}{(1+\zeta)^2-\gamma^2}\right.$$

$$\left.-\frac{1}{\gamma^2}\ln\frac{(1+\zeta)^2}{(1+\zeta)^2-\gamma^2} + \frac{2}{1-\zeta^2+\gamma^2} - \frac{2}{\gamma^2}\ln\frac{1-\zeta^2+\gamma^2}{1-\zeta^2}\right\},$$

(6.74)

where $\zeta = z/d$.

The respective expressions for the force and the stiffness are

$$f_{||} = \frac{F_{||}}{F_0} = \frac{8\gamma^3\zeta\left(3\zeta^4-2\zeta^2-6\zeta^2\gamma^2-1-2\gamma^2+3\gamma^4\right)}{\left(\zeta^2-\gamma^2-1\right)^2\left(1+2\zeta+\zeta^2-\gamma^2\right)^2\left(1-2\zeta+\zeta^2-\gamma^2\right)^2},$$

(6.75)

$$k_{||} = \frac{K_{||}}{K_0}2\gamma^3\left[\frac{5-\gamma^3}{(1-\gamma^2)^3} - \frac{1}{(1+\gamma^2)^2}\right]$$

(6.76)

Expression 6.76 is valid for small and shows that longitudinal stiffness varies with the rotor radius as $1/R^3$. Again, the exact expression is too large to be presented here, but may be obtained in closed form in Maple. Figure 6.7 plots longitudinal and transversal stiffness against γ, longitudinal stiffness is positive for any γ.

Thus, if $\gamma < 0.37$, equilibrium at the point $x = y = z = 0$ is stable. From Figure 6.7, which compares the longitudinal and transverse stiffness, it is seen that the former exceeds the latter and is always positive. Therefore, it is the transverse stiffness that is a critical factor in choosing parameter γ. For opposite charges, equilibrium is unstable, since transverse stiffness $k_\perp$ is negative for any γ.

If the rotor is subjected to gravitational or inertial forces **F** (see Figure 6.8) along the z-axis, the equilibrium position shifts downward and the rotor would sag. The coordinates of a new equilibrium state are found from the expressions:

Force Characteristics of a Magnetic Suspension of an SC Sphere

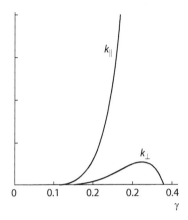

FIGURE 6.7
Longitudinal and transverse stiffness as a function of parameter γ.

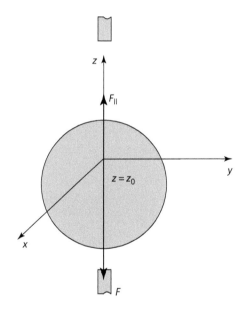

FIGURE 6.8
Spherical sphere in the field of two poles displaced due to gravity.

$$F = F_{\parallel}(z_0), \qquad (6.77)$$

$$mg = \frac{8\gamma^3\zeta\left(3\zeta^4 - 2\zeta^2 - 6\zeta^2\gamma^2 - 1 - 2\gamma^2 + 3\gamma^4\right)}{\left(\zeta^2 - \gamma^2 - 1\right)^2 \left(1 + 2\zeta + \zeta^2 - \gamma^2\right)^2 \left(1 - 2\zeta + \zeta^2 - \gamma^2\right)^2} F_0, \qquad (6.78)$$

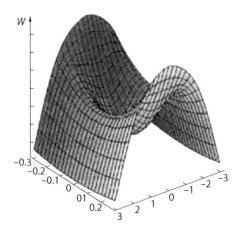

FIGURE 6.9
Dependence of energy on the coordinates z and y.

where g is the acceleration of gravity.

For a given R, d, and q from Equation 6.78, one obtains the equilibrium point z_0; however, the stability of the new equilibrium position must be checked. The longitudinal stiffness is always positive; as to the transverse stability of the new equilibrium state, one must first obtain an expression for the energy at this point and then calculate the stiffness. For this purpose, we will take advantage of the general formula (Equation 6.66) for the energy with $b_1 = \sqrt{y^2 + (d-z)^2}$, $b_2 = \sqrt{y^2 + (d+z)^2}$, $\eta_{12} = 2d$, $x = 0$. The dependence of energy W on the coordinates z and y is shown in Figure 6.9. The second derivative of the energy with respect to y at $z = z_0$ gives the stiffness. It is seen from Figure 6.9 that there is a critical longitudinal displacement, z_{cr}, at which the transverse stability is lost and the transverse stiffness becomes negative. The position of the critical point, z_{cr}, can be determined by solving the following equation in z:

$$\left.\frac{\partial^2 W(y,z)}{\partial y^2}\right|_{y=0} = 0. \tag{6.79}$$

The configuration presented in Figure 6.8 is stable at $z_0 < z_{cr}$. A general solution to Equation 6.79 is impossible to find. Substituting particular values of parameters q, R, and d ($q = 3.5$ A m, $d = 0.03$ m, and $R = 0.01$ m), we find that $z_{cr} = \pm 0.0025$ m. It defines the minimum force leading to laterally unstable equilibrium.

6.6.4 Circular Magnet

As a final example, consider a superconducting sphere in the field of a circular magnet. In the absence of a superconductor, the magnetic field may

Force Characteristics of a Magnetic Suspension of an SC Sphere

be easily calculated. Consider a circular magnet of length $2l$ with an internal radius R_1 and an external radius R_2 (Figure 6.10). We will use again the notion of magnetic charge. A circular magnet may be modeled as a positive and a negative magnetic charge evenly distributed over the top and bottom of the magnet. Magnetic induction **B** may be defined from the Coulomb law. Its components are

$$B_z(\rho,z) = \frac{B_0}{4\pi} \int_{R_1}^{R_2} \int_0^{2\pi} \frac{(z+l)\rho_1 d\rho_1 d\theta_1 dz_1}{\left((z+l)^2 + \rho^2 + \rho_1^2 + 2\rho\rho_1\cos\theta_1\right)^{3/2}}$$

$$- \frac{B_0}{4\pi} \int_{R_1}^{R_2} \int_0^{2\pi} \frac{(z-l)\rho_1 d\rho_1 d\theta_1 dz_1}{\left((z-l)^2 + \rho^2 + \rho_1^2 + 2\rho\rho_1\cos\theta_1\right)^{3/2}},$$ (6.80)

$$B_\rho(\rho,z) = \frac{B_0}{4\pi} \int_{R_1}^{R_2} \int_0^{2\pi} \frac{(\rho - \rho_1\cos\theta_1)\rho_1 d\rho_1 d\theta_1}{\left((z+l^2) + \rho^2 + \rho_1^2 + 2\rho\rho_1\cos\theta_1\right)^{3/2}}$$

$$- \frac{B_0}{4\pi} \int_{R_1}^{R_2} \int_0^{2\pi} \frac{(\rho - \rho_1\cos\theta_1)\rho_1 d\rho_1 d\theta_1}{\left((z-l^2) + \rho^2 + \rho_1^2 + 2\rho\rho_1\cos\theta_1\right)^{3/2}},$$ (6.81)

where:
B_0 is remnant magnetization corresponding to the surface density of the magnetic charge
ρ, z are the coordinates of the point where magnetic induction is calculated
ρ_1, θ_1, z_1 are the integration points

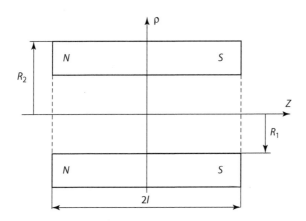

FIGURE 6.10
Circular magnet.

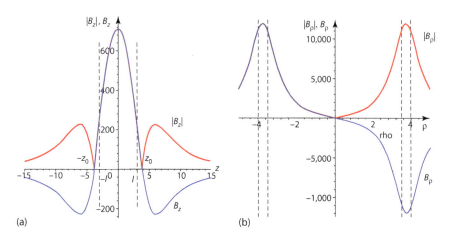

FIGURE 6.11
Axial and radial components of magnetic inductions of a circular magnet as functions of z and ρ, respectively.

The plots showing the axial and radial components of the magnetic induction, created by a circular magnet and their magnitudes (for $R_1 = 3.5$, $R_2 = 4$ cm, $B_0 = 14000 Gs$) are depicted in Figure 6.11a and b, respectively.

The magnets' boundaries are marked by a dashed line. It can be seen that the magnitude of the axial component $|B_z|$ has a minimum near the magnet's surface: the first minimum is located in the point $-z_0 < -l$, the second minimum is located in the point $z_0 > l$. When the length of the magnet l is increased, an additional minimum appears in the point $z = 0$. Thus, the magnitude of the axial component may have one or two minima depending on the length of the magnet. The magnitude of the radial component has a minimum on the magnet's axis. In the points with coordinates $\rho = 0$ and $z = \pm z_0$, quantity $|B|$ reaches minimum and equals 0.

Therefore, in the vicinity of the magnet's surface there exists an MPW, a region with a minimum of magnitude of magnetic induction. With an increase in the magnet's size, this minimum moves toward the plane of the magnet's pole.

When the magnet's length is increased, point z_0 is approaching the point $z = l$. If the magnet is long enough, then one can neglect the second term in Expressions 6.80 and 6.81 and assume that the magnetic field is created by a magnetic ring with a surface charge distributed with surface density B_0. Figure 6.12 shows the components of a magnetic induction generated by a magnetic ring, that is, one pole of a circular magnet (a: transverse component and its magnitude; b: radial component and its magnitude). Here, the center of the magnetic ring coincides with the origin of the coordinates. From Figure 6.12 it can be seen that $|B|$ reaches the plane of the ring on its axis; $|B|$ reaches maximum in the points with coordinates $z = \pm z_m$, and then reduces when $|z| > z_m$.

Force Characteristics of a Magnetic Suspension of an SC Sphere

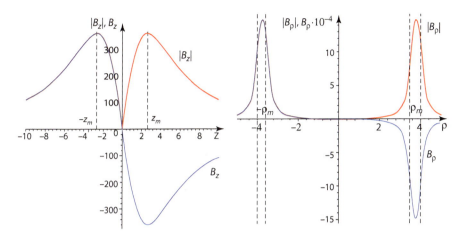

FIGURE 6.12
Axial and radial components of magnetic inductions of a magnetic ring as functions of z and ρ, respectively.

For the stable confinement of a superconducting rotor in the magnetic field, it is necessary that its potential energy reaches minimum. The potential energy of a magnetic field is determined by the square of the magnetic induction. Due to the shape of a circular magnet, a region in the vicinity of the magnet exists where magnetic induction reaches minimum. Thus, we can expect that if we put a diamagnetic in this region, it will experience restoring forces, returning it to this region, and the diamagnetic body may be held both above and below the circular magnet. The latter does not seem possible due to the repulsive character of the magnetic force in a diamagnetic. It is shown [55] that a superconducting sphere may be levitated below a circular magnetic ring.

We will calculate the energy of a spherical superconducting rotor in the field of a thin magnetic ring (Figure 6.13). A magnetic ring is considered to be thin if the difference between the external and internal radiuses is small compared to the internal radius, $\rho_e - \rho_i \ll \rho_i$.

Assuming that the ring is formed by point magnetic charges evenly distributed along the circumference, we can use Expressions 6.33 and 6.62. Given a full charge distributed over the ring, it is easy to calculate the coefficients (Equation 6.62), the dependence of the energy on displacements along the symmetry axis, force, and stiffness.

These calculations were carried out in Maple. Figure 6.14a through c shows the dependencies of energy W, the longitudinal components of force F_z, and the stiffness k_z of the rotor in the field of the circular magnet for values $R = 1$ cm, $\rho_i = 3.95$ cm, $\rho_e = 4.05$ cm, and $q = 280$ A m for displacements along the symmetry axis Oz; $z = 0$ corresponds to the plane of the ring. From Figure 6.14, it can be seen that the stable confinement of the sphere in the

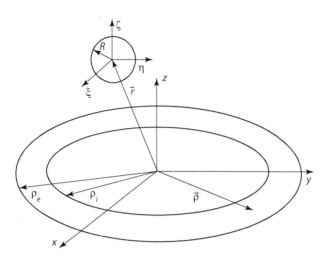

FIGURE 6.13
Superconducting sphere in the field of a magnetic ring.

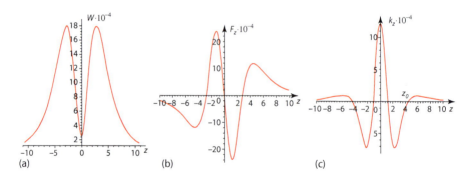

FIGURE 6.14
Dependence of energy, force, and stiffness on the z coordinate.

absence of gravity is possible in the plane of the ring. The calculation of transversal stiffness using Equation 6.45 gives $k_\rho = k_x = k_y = 52,800$ kg/s.

Coefficients Equation 6.62 are calculated only for the displacements along the symmetry axis and are not appropriate for calculating off-axis displacements. More complete information may be obtained from coefficients Equation 6.61. They were used to obtain Expression 6.66, which in the limit transition allows the calculation of energy for distributed sources. After transition from summation to integration, we will have

$$W = \frac{1}{2}\int_{\rho_i}^{\rho_e}\int_{\rho_i}^{\rho_e}\int_0^{2\pi}\int_0^{2\pi} B_0^2 \frac{R\rho_1\rho_2}{b_1b_2} \sum_l \frac{l}{l+1}\left(\frac{R^2}{b_1b_2}\right)^l P_l(\cos\theta_{12})d\varphi_1 d\varphi_2 d\rho_1 d\rho_2. \quad (6.82)$$

Force Characteristics of a Magnetic Suspension of an SC Sphere

The results of calculations using Expression 6.82 are presented in Figure 6.15 for the values $R = 1$ cm, $\rho_i = 3.0$ cm, $\rho_e = 4.0$ cm, and $q = 280$ A·m.

Taking into account the gravity or inertia force along Oz would cause a shift of the equilibrium position to the point z_e (Figure 6.16), which is determined from the equilibrium condition:

$$F_g - \frac{dW(z)}{dz}\bigg|_{z=z_e} = 0. \tag{6.83}$$

If stiffness is positive in this point, that is, condition $|z_e| < |z_0|$ is satisfied, then a new equilibrium state will be stable. From Figure 6.16, one can see that the stable confinement of a rotor is possible both above and below the magnet, despite the repulsive character of the magnetic force. Forces from the first magnet previously considered are reaching 1 N and allow the holding of an in-field mass up to 100 g.

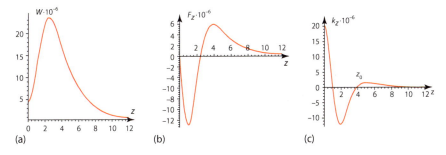

FIGURE 6.15
Energy, axial force, stiffness, and functions of the z coordinate calculated using Expression 6.82.

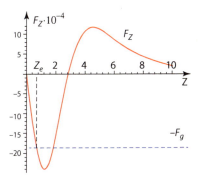

FIGURE 6.16
Magnetic force, force of gravity, and equilibrium position.

6.7 Conclusions

We presented a model to calculate the interaction energy between a superconducting sphere and a supporting magnetic field generated by a system of magnetic charges or current loops. In some important cases, these sums may be converted to simple closed-form analytical expressions. This approach allows us to consider different configurations of sources, to find lifting forces and stiffness, and to look for the source distributions providing the desired force characteristics of the suspension, and provides a means to analytically study dynamics and a superconducting spherical and close to spherical body in the field generated by a system of magnetic charges or coaxial current loops.

References

1. J. R. Hull, Using high-temperature superconductors for levitation applications, *JOM*, 51(7), pp. 13–18, 1999.
2. K. B. Ma, Y. V. Postrekhin, and K. K. Chu, Superconductor and magnetic levitation devices, *Review of Scientific Instruments*, 74(12), pp. 4989–5017, 2003.
3. F. C. Moon, *Superconducting Levitation*, Wiley, Weinheim, 2004.
4. R. Moser, J. Sandtner, and H. Bleuler, Diamagnetic suspension system for small rotors, *Journal of Mechatronics*, 1(2), pp. 131–137, 2001.
5. L. Rossini, O. Chetelat, E. Onillon, and Y. Perriad, Force and torque analytical models of a reaction sphere actuator based on spherical harmonic rotation and decomposition, *IIIE/ASME Transactions on Mechatronics*, 18(3), pp. 1006–1018, 2013.
6. G. C. Schroll, Design of a spherical vehicle with flywheel momentum storage for high torque capabilities, Thesis for Bachelor Degree, MIT, 2008.
7. B. P. Mann and N. D. Sims, Energy harvesting from the nonlinear oscillations of magnetic levitation, *Journal of Sound and Vibration*, 319, pp. 515–530, 2009.
8. M. E. Hoque and T. Mizuno, Magnetic levitation technique for active vibration control, in B. Polajzer (Ed.), *Magnetic Bearings, Theory and Applications*, Chapter 4, InTech, 2010.
9. Y. Zhu, Q. Li, D. Xu, C. Hu, and M. Zhang, Modeling and analysis of a negative stiffness magnetic suspension vibrator isolator with experimental investigations, *Review of Scientific Instruments*, 93, 095108, 2012.
10. J. Szekely, E. Schwartz, and R. Hyers, Electromagnetic levitation: A useful tool in microgravity research, *JOM*, 47(5), pp. 50–53, 1995.
11. M. J. A. Moes, J. C. Gielen, R.-J. Bleichrodt, J. W. A. van Loon, P. C. M. Christianen, and J. Boonstra, Simulation of microgravity by magnetic levitation and random positioning: Effect on human A431 cell morphology, *Microgravity Science and Technology*, 23(2), pp. 249–261, 2011.
12. B. E. Hammer, L. S. Kidder, P. C. Williams, and W. W. Xu, Magnetic levitation of MC3T3 osteoblast cells as a ground-based simulation of microgravity, *Microgravity Science and Technology*, 21(4), pp. 311–318, 2009.

13. C. E. Dijkstra et al., Diamagnetic levitation enhances growth of liquid bacterials cultures by increasing oxygen availability, *Journal of the Royal Society Interface*, 8(56), pp. 334–344, 2010.
14. H. Minagawa et al., Control of levitation in electromagnetic levitators under microgravity, *Japanese Journal of Applied Physics*, 35(Pt. 2 No. 12B, L1714), 1996.
15. H. Chetouani, C. Jeandey, V. Haguet, H. Rostaing, C. Dieppedale, and G. Reyne, Diamagnetic levitation with permanent magnets for contactless guiding and trapping of microdroplets and particles in air and liquids, *IEEE Transactions on Magnetics* 42(10), pp. 3557–3559, 2006.
16. P. Kaufmann, P. Pham, A. Masse, M. Kustov, T. Honegger, D. Peyrade, V. Haguet, and G. Reyne, Contactless dielectrophoretic handling of diamagnetic levitating water droplets in air, *IEEE Transactions on Magnetics*, 46(8), p. 201, 2006.
17. J. Castillo, M. Dimaki, and W. E. Svendsen, Manipulation of biological samples using micro and nano techniques, *Integrative Biology*, 1, pp. 30–42, 2009.
18. T. Inamori and S. Nakasuka, Application of magnetic sensors to nano and micro-satellite attitude control systems, in K. Kuang (Ed.), *Magnetic Sensors: Principles and Applications*, InTech, 2012, pp. 85–102.
19. W. Braunbeck, Freischwebende Korper im elektischen und magnetischen Feld, *Z. Phys.* 112, p. 753, 1939.
20. H. Austin and K. T. McDonald, *Diamagnetic Levitation*, Joseph Henry Laboratories, Princeton University Communication, Nov. 15, 2001.
21. R. Perline, Diamagnetic levitation, *American Scientist*, 92(5), pp. 428–435, 2004.
22. M. D. Simon and A. K. Geim, Diamagnetic levitation: Flying frogs and floating magnets, *Journal of Applied Physics*, 87(9), pp. 6200–6204, 2000.
23. Y. M. Urman et al., On levitation of diamagnetic bodies in a magnetic field, *Technical Physics*, 55(9), pp. 1257–1265, 2010.
24. M. D. Simon, L. O. Heflinger, and A. K. Geim, Diamagnetically stabilized magnet levitation, *American Journal of Physics*, 69(6), pp. 702–713, 2001.
25. V. S. Mikhalevich, V. V. Kozorez, V. M. Rashkovan, et al., *Magnetic Potential Well: The Effect of Stabilization of Dynamic Superconducting Systems*, Naukova Dumka, Kiev, 1991 [in Russian].
26. S. Kuznetsov, and J. K. Guest, Topology optimization of magnetic source distributions for diamagnetic and superconducting levitation, *Journal of Magnetism and Magnetic Materials*, 438, pp. 60–69, 2017.
27. Y.M. Urman, and V. V. Novikov, Effect of elasticity on the dynamics of a superconducting rotor rotating in a magnetic field, *Journal of Applied Mechanics and Technical Physics*, 42(3), pp. 518–522, 2001.
28. M. V. Berry and A. K. Geim, Of flying frogs and levitrons, *European Journal of Physics*, 18, pp. 307–313, 1997.
29. S. I. Kuznetsov and Y. M. Urman, Levitation of a superconducting body in the field of N magnetic poles, *Technical Physics*, 51, pp. 297–306, 2006.
30. S. I. Kuznetsov, A. O. Malkhanov, and Yu. M. Urman, Effect of periodic shape changes in the shape of a superconducting body on its dynamics in a noncontact magnetic suspension, *Technical Physics*, 53(12), pp. 1529–1535, 2008.
31. Y. M. Xiao et al., Observation of the London moment and trapped flux in precision gyroscopes, *IEEE Transactions on Applied Superconductivity*, 3(1), pp. 2144, 1993.
32. Y.M. Urman, Method of irreducible tensors in problems of evolution motions of a rigid body with a fixed point, *Mechanics of Solids*, 4, pp. 10–20, 1997.

33. Y.M. Urman, Method of irreducible tensors in problems of rotation of a conductive body in nonuniform magnetic fields, *Mechanics of Solids*, 36(2), pp. 8–15, 2001.
34. Y.M. Urman, Irreducible tensors and their application in problems of dynamics of solids, *Mechanics of Solids*, 42(6), pp. 52–68, 2007.
35. J. D. Jackson, *Classical Electrodynamics*, 3rd ed., Wiley, New York, 1999.
36. I. V. Veselitskii and Y. M. Urman, Integral representation of inductance for the system superconducting sphere-current coils, *Sov. Phys. Tech. Phys.*, 24, 881, 1979.
37. Y.M. Urman, Theory for the calculation of the force characteristics of an electromagnetic suspension of a superconducting body, *Technical Physics*, 42(1), pp. 1–6, 1997.
38. Y.M. Urman, Calculation of the force characteristics of a multi-coil suspension of a superconducting sphere, *Technical Physics*, 42(1), pp. 7–13, 1997.
39. V. E. Plehanov, A. I. Chernomorkiy, and K. B. Yakovlev, Numerical calculations of spherical contactless suspension, in A.V. Remnikov (Ed.), *Systems for Orientation and Navigation and Their Elements*, Moscow Aviation Institute, Moscow, 1979, pp. 118–122 [in Russian].
40. S. A. Kostrykov, V. V. Peshkov, and G. E. Shunin, Finite elements method in the computer modeling of superconducting screens and suspensions, *Bulletin of the Russian Academy of Sciences: Physics*, 61(5), pp. 985–989.
41. G. E. Shunin et al., FEMPDESolver 2.0 software package for finite-element analysis of superconducting current systems, *Bulletin of the Russian Academy of Sciences: Physics*, 68(7), pp. 1172–1180, 2004
42. Y. S. Ermolaev and I. A. Rudnev, A method to calculate levitation force in the system magnet-superconductor, *Letters to Technical Physics*, 31(24), pp. 60–66, 2005
43. L. I. Bataronov, S. A. Kostrykov, V. V. Peshkov, and G. E. Shunin, Computer modeling of spherical superconducting suspension,, *Bulletin of the Russian Academy of Sciences: Physics*, 70(8), 2006
44. M. I. Bataronova, S. A. Kostrykov, V. V. Peshkov, and G. E. Shunin, Computer modeling of superconducting systems using finite elements method, *Bulletin of the Voronezh State Technical University*, 3(8), 2007
45. G. E. Shunin, S. A. Kostrykov, V. V. Peshkov, and M. I. Islentieva, The development of the system of superconducting suspension computer simulation, *Bulletin of Voronezh State Technical University*, 10(1), pp. 21–26, 2014
46. L. D. Landau and E. M. Lifshitz, *Electrodynamics of Continuous Media* (Volume 8 of A Course of Theoretical Physics), Butterworth and Heinemann, Oxford, 1984.
47. A. N. Tverdokhlebov and A. L. Shuster, A sphere in an arbitrary quasistatic electric or magnetic field, *Journal of Technical Physics*, 42(9), pp. 2477–2484, 1972 [in Russian].
48. D. A. Varshalovich et al., *Quantum Theory of Angular Momentum*, World Scientific, Singapore, 1988.
49. Yu. M. Urman, To the Calculation of Force Characteristics of External Spherical Suspension of Cryogenic Gyroscope, Izv. Vyssh. Uchebn. Zaved., *Priborostroenie*, 8(72), pp. 72–74, 1973 [in Russian].
50. D. Skubov and K. Khodzaev, *Non-Linear Electromechanics*, Springer, Berlin Heidelberg, 2008.

51. L. D. Landau and E. M. Lifshitz, *Mechanics* (Volume 1 of A Course of Theoretical Physics), Butterworth-Heinemann, Oxford, 1976.
52. K. P. Belov, *Rare-Earth Magnets and Their Applications*, Nauka, Moscow, 1980 [in Russian].
53. Pyatin, Yu. M. (Ed.), *Permanent Magnets: A Handbook*, Energiya, Moscow, 1980 [in Russian].
54. W. R. Smythe, *Static and Dynamic Electricity*, 2nd ed., McGraw-Hill, New York, 1950.
55. S. I. Kuznetsov and Yu. M. Urman, On the possibility of levitation of superconducting sphere in the field of circular magnet, *Bulletin of N. I. Lobachevsky State University of Nizhny Novgorod*. Series: Mechanics, 1(7), pp. 5–14, 2006.

7

Magnetic Angle Sensors

Udo Ausserlechner

CONTENTS

7.1 Scope of This Chapter 202
7.2 Technologies of Magnetic Field Sensing Elements 203
    7.2.1 Magnetoresistors 204
    7.2.2 Hall Effect Devices 209
    7.2.3 Comparison of Technologies for Magnetic Angle Sensing Devices 211
7.3 Assembly Tolerances and Sensor Offset, Gain Mismatch, and Misalignment 216
    7.3.1 Assembly Tolerances, Offset, Gain Mismatch, and Misalignment in GMRs/TMRs 224
        7.3.1.1 Autocalibration of GMR Angle Sensors for Continuously Rotating Shafts 231
    7.3.2 Assembly Tolerances, Offset, Gain Mismatch, and Misalignment in AMRs, VHalls, and HHalls with IMC 236
7.4 Field-Gradient Angle Sensors 237
7.5 Magnets for Magnetic Angle Sensors 241
References 246

ABSTRACT: For many industrial and automotive applications, one needs to measure the absolute rotational position of a shaft with an accuracy of about 1°, a resolution of 0.1°, and costs of a few euros. Examples are headlight leveling (120° stroke), throttle valve position (120° stroke), motor commutation and torque control (360° range), cam shaft and cam phaser angle sensing, steering wheel (10*360° range), and rotational vibration damping (20*360° range) [1]. Although some of these applications can also be served by potentiometers and optical, capacitive, and inductive angle sensors, a major group of applications calls for magnetic angle sensors [2]. They are small, their installation requires only moderate assembly tolerances, they are free of wear, they are robust against dirt and pollution, they are also sufficiently robust against electromagnetic interference, they are fast, and they are economic.

7.1 Scope of This Chapter

For many industrial and automotive applications, one needs to measure the absolute rotational position of a shaft with an accuracy of about 1°, a resolution of 0.1°, and costs of a few euros. Examples are headlight leveling (120° stroke), throttle valve position (120° stroke), motor commutation and torque control (360° range), cam shaft and cam phaser angle sensing, steering wheel (10*360° range), and rotational vibration damping (20*360° range) [1]. Although some of these applications can also be served by potentiometers and optical, capacitive, and inductive angle sensors, a major group of applications calls for magnetic angle sensors [2]. They are small, their installation requires only moderate assembly tolerances, they are free of wear, they are robust against dirt and pollution, they are also sufficiently robust against electromagnetic interference, they are fast, and they are economic. This is the type of angle sensor that we will discuss in the following. We limit the discussion to the position detection of a shaft where magnetic fields of the order of 40 mT are used. Hence, weak-field sensors, such as compass sensors [3] and navigation via magnetic fields [4], and very small angle strokes as in torque meters [5] are excluded. The latter uses large ferrous structures that collect magnetic flux from a permanent magnet depending on the rotational position and guide this flux to a magnetic field sensor (e.g., Hall plate). Some throttle valve position sensors use similar macroscopic flux guides to measure angles with strokes of 100°–160° [2]. We exclude such sensors with macroscopic flux guides from our discussion. We also skip the discussion of incremental angle sensors that are in widespread use in anti-lock braking systems (ABS) and for transmission-, crank-, and cam-angle sensors [6,7].

We consider angle sensors according to Figure 7.1: they have a permanent magnet attached to a rotatable shaft. Through-shaft sensors have a long shaft whose ends are not accessible to the sensor. Thus, a ring magnet has to be slipped over the shaft. If an end of the shaft can be used to mount a magnet, we call this an *end-of-shaft sensor*. In general, the best performance is obtained if we drill a hole into the end of the shaft to accommodate the magnet—this is called an *in-shaft sensor*. End-of-shaft sensors may have

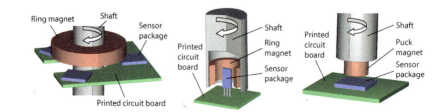

FIGURE 7.1
Three types of magnetic angle sensors: through-shaft, in-shaft, end-of-shaft. Parts of the in-shaft sensor are cut off to give a view of the sensor package.

Magnetic Angle Sensors

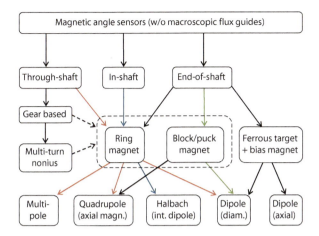

FIGURE 7.2
Classification scheme of common magnetic angle sensor systems.

small, simple, and cheap magnets in the shape of blocks (parallelepipeds) or short cylinders (like a puck). In-shaft sensors preferably use small ring magnets, and through-shaft sensors need rather large ring magnets, which are more expensive to manufacture. Figure 7.2 gives a classification scheme of these common magnetic angle sensors.

In most practical cases, the shaft is ferrous; a soft-magnetic material with a large relative permeability of $\mu_r > 100$, typically on the order of 1000–2500. Therefore, the shaft interacts with the magnet and distorts its magnetic field. Moreover, the shaft also interacts with external magnetic disturbance fields, which can be an advantage or a drawback.

End-of-shaft and in-shaft sensors may use single sensor packages with single sensor chips, because they need to sample the magnetic field only in a single spot right on the rotation axis or in the immediate vicinity (±2 mm) around the rotation axis. Through-shaft sensors usually need two, three, or more sensor packages because they have to measure the field on a concentric reading circle around the shaft. Thus, through-shaft sensors are more challenging as they have larger magnets and more sensor packages and these increase the system's costs. Due to space limitations, we will only touch on them here.

7.2 Technologies of Magnetic Field Sensing Elements

The rotating magnetic field has to be detected by magnetic field sensing elements. There are several competing sensor technologies such as various types of magnetoresistors (MR) and Hall effect devices (Figure 7.3). Often, commercial suppliers specialize in a specific technology and tend to

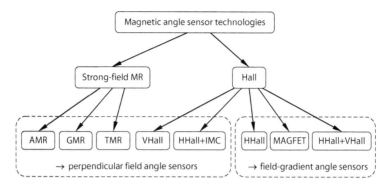

FIGURE 7.3
The most common sensor technologies used for magnetic angle sensors.

overemphasize the advantages of their proprietary one. Only a few companies are in the luxurious position to offer many kinds of sensor technologies. With the experience he has gained in one such company, the author has come to the conclusion that the differences in the various sensor technologies for angle sensors matter less than generally presumed. This is sketched in the subsequent introduction to these sensor technologies followed by a comparison of their pros and cons with respect to angle sensors.

7.2.1 Magnetoresistors

Magnetoresistance (MR) is generally defined as the change of electrical resistance in a material caused by a magnetic field.

$$\text{MR} = \frac{R_{\max} - R_{\min}}{R_{\min}} \quad (7.1)$$

$R_{\max}$ and $R_{\min}$ are the maximum and minimum observed resistances, respectively. The anisotropic MR (AMR) was first observed in 1857 [8]. It takes place in a thin sheet—the free layer—made of a ferromagnetic metal such as Ni, Co, or Fe. In practice, one prefers permalloy, an alloy of 81% Ni and 19% Fe, because there the magnetostrictive constant vanishes and this makes the resistance independent of mechanical stress. In the absence of an applied magnetic field, the magnetization in the thin free layer should be homogeneous ("single domain particle") and it is aligned along an easy axis. The easy axis is defined by the geometry of the device via the so-called shape anisotropy: in permalloy it is along the longest dimension of the sample. MRs are very thin (~30 nm). Their thickness is much lower than the lateral (x,y)-dimensions and so the lateral dimensions define the sensitive plane, which is identical to the main surface of the substrate to which the AMRs are sputtered and which we call the (x,y)-plane. As a consequence of shape anisotropy, their magnetization is parallel to the (x,y)-plane and the applied magnetic field only rotates the magnetization in this (x,y)-plane. Therefore,

Magnetic Angle Sensors

MRs respond only to the in-plane component of the applied magnetic field $\vec{B}_{\text{in-plane}} = \vec{B} - (\vec{B} \cdot \vec{n}_z)\vec{n}_z = B_x \vec{n}_x + B_y \vec{n}_y$. If an external magnetic field is applied to the MR and points in a direction different from the easy axis, it tries to pull out the magnetization from its easy axis. In *strong-field angle sensors*, the field is much stronger than the lateral shape anisotropy so that the magnetization points essentially in the same direction as the applied field. Then, the resistance of the free layer depends on the angle φ between the applied in-plane field component and the current streamlines according to

$$R^{(\text{AMR})} = R_{\min}\left\{1 + MR^{(\text{AMR})}(\cos\varphi)^2\right\}$$

$$= \frac{R_{\max} + R_{\min}}{2}\left\{1 + \frac{R_{\max} - R_{\min}}{R_{\max} + R_{\min}}\cos(2\varphi)\right\} = R_{\min}\frac{1 + \frac{h}{2}\cos(2\varphi)}{1 - h/2} \quad (7.2)$$

So, the resistance is at maximum when the magnetization direction is parallel to the current flow and it is at minimum when the magnetization direction is orthogonal to the current flow. The resistance does not change if one swaps the polarity of the current and/or the magnetic field—the φ-periodicity is 180°. Therefore, simple* AMR angle sensors cannot discriminate rotational positions φ against φ + 180°—they are unambiguous only within an angular stroke of 180°. In the following, we replace $(R_{\max} - R_{\min})/(R_{\max} + R_{\min}) \to h/2$. It holds that $h = 2MR/(2 + MR)$ and we call this the *magnetic stroke* or the *magnetic sensitivity*. The AMR effect is $MR^{(\text{AMR})} = 0.025 \cong h$ for a 30 nm thick permalloy film and it decreases for thinner layers [10]. In commercial devices, layout restrictions and wiring reduce this value by about 10%.

Giant MR (GMR) was discovered in the mid-1980s. In GMRs, several ferromagnetic regions (Co or Fe) with different directions of magnetization are closer together than a certain critical length (the mean free path of the spin of the conduction electrons or their spin diffusion length). Therefore, the electrons do not change their spin when they traverse these regions, which are usually thin layers. Each time an electron crosses a region with magnetization anti-parallel to its spin, it is scattered much more than if the magnetization were parallel to its spin. More scattering means higher electrical resistance. In multilayer GMRs, several ferromagnetic layers are magnetized with alternating polarity at a zero applied magnetic field. Therefore, spin-up and spin-down electrons are scattered in every second layer. If a sufficiently strong magnetic field is applied, it pulls the magnetizations of all layers in the same direction. Then, the electrons

---

* Some commercial suppliers offer advanced sensor systems where integrated coils superimpose additional magnetic fields on the AMRs [9]. From the changes of signals induced by these superimposed fields, they can discriminate rotational positions φ against φ + 180°. Other systems add Hall sensors to discriminate φ against φ + 180°. Their intention is to use AMR technology to measure full 360° rotations, because in the past AMR technology achieved the highest angle accuracy at the highest bandwidth.

with their spin parallel to this common direction have no scattering while the other electrons are scattered in each ferromagnetic layer, and this reduces the overall resistance. The alternating direction of magnetization in neighboring layers at zero external field can be obtained by non-magnetic spacer layers in-between (Au, Cr, Cu). The thickness of these spacer layers determines if the two magnetization directions are parallel or anti-parallel at zero field [11]. The strength of this ferromagnetic or anti-ferromagnetic coupling can be varied via thickness according to theory (oscillatory RKKY exchange coupling). For specific thicknesses of the spacer layer, the coupling between neighboring ferromagnetic layers even vanishes. This is used in spin valve angle GMRs, where one ferromagnetic layer ("hard" or "pinned" or "reference" layer) has a fixed reference magnetization while the magnetization of the other layer is free to follow the applied magnetic field ("soft" or "free" or "sensing" layer) [12]. The direction of magnetization of the pinned layer is defined by a so-called exchange bias coupling to an anti-ferromagnet, which usually comprises one layer next to the pinned layer. The anti-ferromagnet is magnetized by a strong external field (in the range of 1 T) at an elevated temperature (>200°C) during the fabrication of the GMR. Spin valve GMRs have fewer layers and therefore they are cheaper to manufacture but they have lower MR ratios than multilayer GMRs. With spin-valve GMRs, one can make sensors, which are sensitive to the strength of an applied field (so-called weak-field sensors) or to the direction of an applied field (so-called strong-field or saturated sensors):

1. In weak-field sensors, the free layer is designed to have an anisotropy perpendicular to the direction of the pinned layer magnetization. This means that at zero external field, the magnetization of the free layer is at a right angle (90°) to the magnetization of the pinned layer. The anisotropy can be defined by the elongated shape of the free layer, or by its growth conditions, or by an anneal in a magnetic field. Then, for small applied fields, the resistance varies linearly around zero field. For larger applied fields, the magnetization of the free layer is parallel (+Hk) or anti-parallel (−Hk) to the pinned layer. This type of GMR is not used for the types of magnetic angle sensors that we are discussing in this chapter.

2. In strong-field sensors, the free layer is designed to have zero anisotropy so that its magnetization can freely follow a rotating applied field. On the other hand, it is necessary for the applied field to be continuously present during operation, otherwise the magnetization of the free layer is undefined. If the applied field is present, but too weak, the magnetization of the free layer may not be perfectly parallel to the applied field, because it may be trapped by small imperfections. This will lead to hysteresis when the rotation direction is reversed.

In GMRs, the non-magnetic spacer is a metal while in tunneling MRs (TMRs) it is a dielectric sufficiently thin for tunneling current. Due to the

Magnetic Angle Sensors

tunneling effect, TMRs have different temperature coefficients and they are electrically nonlinear (therefore, they are often called magnetic tunnel junctions [MTJs] instead of resistors). They have their maximum sensitivity at low voltage (<100 mV). Therefore, one has to connect many elements in series for common supply voltages of several volts. Since the tunnel junction generates notable excess noise, one may reduce the effective resistance by the parallel connection of several elements. In the end, this leads to TMR angle sensors, which are not much smaller than AMR/GMR angle sensors. In a spin-valve GMR/TMR, the resistance changes as a function of angle φ between the magnetizations of free and pinned layers [13], whereby in strong-field angle GMR/TMRs the free layer magnetization very accurately points in the same direction as the applied magnetic in-plane field.

$$R^{(GMR)} = R_{min}\left\{1 + MR^{(GMR)}\frac{1-\cos\varphi}{2}\right\}$$

$$= \frac{R_{max} + R_{min}}{2}\left\{1 - \frac{R_{max} - R_{min}}{R_{max} + R_{min}}\cos\varphi\right\} = R_{min}\frac{1 - \frac{h}{2}\cos\varphi}{1 - h/2} \quad (7.3a)$$

The literature also states that in TMRs the conductance instead of the resistance varies with $\cos\varphi$ [13].

$$R^{(TMR)} = R_{min}\frac{1 + MR^{(TMR)}}{1 + MR^{(TMR)}\frac{1+\cos\varphi}{2}}$$

$$= \frac{2R_{max}R_{min}}{R_{max} + R_{min}}\frac{1}{1 + \frac{R_{max}-R_{min}}{R_{max}+R_{min}}\cos\varphi} = R_{min}\frac{1+h/2}{1+\frac{h}{2}\cos\varphi} \quad (7.3b)$$

For angle sensors, the difference between Equations 7.3a and 7.3b is irrelevant, because both cases give the same output signal of a typical Wheatstone bridge as in Figure 7.6. In both cases, the resistance is at a minimum when the magnetization vectors in the free layer and in the pinned layer are parallel ($\varphi = 0°$). The resistance is at a maximum when the magnetization vectors in the free layer and in the pinned layer are anti-parallel ($\varphi = 180°$). The resistance of a single GMR/TMR has a 360° periodicity. Therefore, GMR/TMR angle sensors can measure the direction of an applied magnetic field unambiguously within an angular stroke of 360°. A typical sensitivity for strong-field angle GMRs is $MR^{(GMR)} = 0.04 \cong h$ and for strong-field angle TMRs a value of $MR^{(TMR)} = 0.86$ ($h = 0.6$) is reported in [14]. Hence, the output signals of AMR, GMR, and TMR scale as 1:1.8:27, but the angle resolution of TMRs is only approximately 10 times better than for AMRs (cf. Table 7.1). The AMR effect is also present in GMR/TMR devices and should be eliminated in practice [15,16].

TABLE 7.1
Comparison of "Ex Fab" (i.e., Uncalibrated) Key Parameters of Common Magnetic Angle Sensor Technologies

	AMR	GMR	TMR	HHall	HHall with IMC	VHall
Angle range	180°	360°	360°	360°	360°	360°
Magnetic sensitivity	$h = 0.022$	$h = 0.04$	$h = 0.6$	$S_u = 51$ mV/V/T	$S_u = 255$ mV/V/T	$S_u = 40$ mV/V/T
Peak **signal amplitude** at 2 V supply	22 mV	40 mV	600 mV	4.6 mV for $B = 45$ mT	23 mV for $B = 45$ mT	3.6 mV for $B = 45$ mT
Impedance per sin/cos-channel	1 kOhm	1 kOhm	1 kOhm	1 kOhm	1 kOhm	1 kOhm
Thermal noise (rms) at 300 K per channel	4.07 nV/sqrt (Hz)	4.07 nV/sqrt (Hz)	4.07 nV/sqrt (Hz)	4.07 nV/sqrt (Hz)	4.07 nV/sqrt (Hz)	4.07 nV/sqrt (Hz)
Flicker noise	x	x	xx(!)	0	0	0
Standard deviation of **angle noise** for 150 kHz corner frequency of first-order low-pass filter including signal conditioning (NF = 1)	0.0036° +x	0.0040° +x	0.0003° + xx(!)	0.0349°	0.0070°	0.0445°
Standard deviation of **signal offset** per channel	0.33 mV	0.67 mV	1.7 mV	30 μT	30 μT	100 μT…30 μT (stacked+parallel)
Standard deviation of **angle error due to signal offset**	0.43°	0.95°	0.16°	0.04°	0.04°	0.14°…0.04° (stacked+parallel)
Worst-case (=6 sigma) **gain mismatch error** of sin/cos-channels	0.5% =0.07°	3% =0.85°	1% =0.29°	1% =0.29°	2% =0.57°	1% =0.29°
Worst-case (=6 sigma) **orthogonality error** of sin/cos-channels	0°	3°	2°	0°	3°	0.5°
Hysteresis error	0.05°	0.05°	0.05°	0°	0.05°	0°
Worst-case **anisotropy/nonlinearity errors** over temperature and lifetime	0.11°	0.4°	0.3°	<0.01°	<0.1°	<0.01°

Note: The values given do not refer to real products, but denote sound limits that can be achieved by optimized design with state-of-the-art technology.

7.2.2 Hall Effect Devices

The discovery of the Hall effect dates back nearly as far as the AMR effect [17]. Semiconductor Hall plates have been widely used in industrial applications since the 1950s and they have been integrated into silicon application-specific integrated circuits (ASICs) since the 1970s. In the 1990s, complementary metal-oxide semiconductor (CMOS) technology was mature enough to implement the spinning current principle on-chip [18], which greatly improved the initial or raw offset of silicon Hall plates of around 2–5 mT (standard deviation) down to the residual offset of spinning current Hall plates of around 5–50 µT (standard deviation). Finally, in the last decade, the long-term drift of their magnetic sensitivity induced by mechanical stress was pushed well below 1% by stress compensation circuits [19]. Hence, silicon Hall plates are a very well understood technology.

In silicon technology, Hall plates are simple tubs with low n-doping, roughly $50 \times 50 \times 1$ µm large, with four contacts regularly spaced around the perimeter of the tub. Electric current flows through two diametrically opposite contacts and the voltage is tapped between the other two contacts. The symmetrical placement of the output contacts ideally gives no voltage between them in the absence of a magnetic field. If a magnetic field is applied perpendicular to the Hall plate, a Lorentz force acts on the majority carriers in the Hall plate and pushes them closer to one contact. This gives an asymmetrical potential in the Hall plate and an output voltage $V_{out,h}$ between the two output contacts, which is fairly linear to the applied field:

$$V_{out,h} = B_z S_u V_{in,h} \tag{7.4a}$$

with the magnetic flux density B_z orthogonal through the Hall plate, the voltage-related magnetic sensitivity S_u, and the supply voltage $V_{in,h}$ across the Hall plate. The output voltage reverses its sign when the magnetic field reverses its sign. Therefore, Hall effect devices can detect rotational positions unambiguously in the full range of 360°. Their magnetic sensitivity, S_u, is

$$S_u = \mu_{n,h} \frac{G_H}{(L/W)_{eff}} \tag{7.4b}$$

with $\mu_{n,h}$ the Hall mobility of electrons in low n-doped silicon, G_H, the Hall geometry factor that describes the short-circuiting effects of finite sized contacts, and $(L/W)_{eff}$ the effective number of squares (which is the ratio of input resistance over sheet resistance). These quantities can be computed and measured very accurately [20–22]. In practice, one wants to maximize the signal-to-noise ratio (SNR) of a Hall plate at a fixed input resistance, which is suitable for the signal conditioning circuit. Thereby, the noise is merely thermal noise, because low-frequency noise is cancelled out by the spinning current Hall probe technique [23]. It can be shown that this is equivalent to maximizing S_u while keeping the input resistance constant. This calls for high mobility and thus low doping according to Equation 7.4b. In silicon, a low phosphorus doping of $2 \times 10^{16}/\text{cm}^3$ gives a Hall mobility of 0.1084/T. The second term on the right-hand side of Equation 7.4b depends only on the layout. For Hall plates with 90° symmetry and at a moderate magnetic field, it is maximized by medium-sized contacts with $(L/W)_{\text{eff}} = \sqrt{2}$, from which follows $G_H = 2/3$ (see appendix B of [24], where it is also shown that Hall plates without 90° symmetry have smaller SNR at the same supply voltage and current). So, the maximum SNR of Hall plates for fixed input resistance is obtained for $S_u \cong 0.108 \times \sqrt{2}/3 \cong 51\,\text{mV/V/T}$. For larger contacts, larger doping, and thicker Hall plates, it is possible to further increase the SNR at a fixed supply voltage, but this drastically increases current consumption.

In practice, the circuit design engineer chooses the lowest Hall plate resistance with which the sensor can still fulfill the specification of maximum allowed current drain. Then, he or she opts for the lowest Hall doping level that Bi/CMOS technology allows. He or she chooses the layout in order to fulfill $(L/W)_{\text{eff}} = \sqrt{2}$. This defines the Hall plate thickness in order to achieve the target resistance. If this Hall plate thickness is not available in the given Bi/CMOS technology, one can connect various Hall plates in parallel to replace a single thick Hall plate, but this needs more chip area.

A major error of Hall effect devices is offset: due to tiny unavoidable asymmetries in the device, the output voltage is not exactly zero at a vanishing magnetic field. However, the Hall sensor circuit operates the Hall effect device in several consecutive operating phases, whereby inputs and outputs are swapped and the outputs of these phases are added up [18]. This is known as the *spinning current Hall probe* and it drastically reduces the offset error from an equivalent value of several millitesla (mT) down to several tens of microtesla (μT), in some cases even down to 1 μT [3]. It also eliminates flicker noise. The operating phases are executed at frequencies up to several hundred kilohertz (kHz), so that analog bandwidths of 150 kHz are achievable, and this is sufficient for nearly all angle sensing applications. Note that the scheme of the spinning current Hall probe eliminates Hall plate offsets without knowing the exact value of the offset, whereas the offset of MRs can only be reduced by first measuring it, storing it as a calibration data in a memory, and subtracting this value from the signals during operation in the field.* Therefore, Hall effect devices need no individual offset calibration

---

* Here we talk of offset cancellation at a static magnetic field. If the magnetic field varies sinusoidally due to a continuous rotation of the magnet in an angle sensor, MR sensors can cancel out major parts of the offset by autocalibration schemes, which are discussed in Section 7.3.1. But this is a completely different procedure, which only works at the system level and requires continuous 360° rotation of the magnetic field. Such an autocalibration scheme works for all types of MR sensors, and it works also for Hall sensors. In Hall sensors, autocalibration can be used on top of spinning current schemes to suppress offset errors below conventional detection limits.

Magnetic Angle Sensors

as MR sensors do, and this reduces the test costs, temperature, and lifetime drifts of Hall effect devices.

Magnetic field-effect transistors (MAG-FETs) are N-Channel MOS (NMOS) transistors with additional sense contacts. In total, they may have three (split-drain MAG-FET) or four contacts (like Hall plates). MAG-FETs use the Hall effect in their thin channel region, and therefore they can be optimized for low current consumption. The spinning current principle also works with MAG-FETs, but due to their electrical nonlinearity it is less efficient than for Hall plates.

The basic idea behind vertical Hall effect devices (VHalls) is to measure a magnetic field in the x-direction by turning a Hall plate from its horizontal (x,y)-plane into its vertical (y,z)-plane. This was originally proposed in 1984 [25]. The first obvious problem of VHalls in Bi/CMOS technology is that it is difficult to make electrical contacts on the three sides of the vertical plate, which are in the depth of the substrate. Yet, it is possible to make contacts on a single side of a Hall plate [26]. This reduces the electrical symmetry of the device, but in linear theory the spinning current scheme still works perfectly for asymmetric devices [27]. The real blocking point for VHalls is the low depths of 3–10 µm of available tubs in Bi/CMOS technologies, because the spacing of the contacts on the top side has to be roughly equal to the depth to get a strong signal. Therefore, the contact spacing of VHalls is only a few micrometers, whereas in traditional Hall plates it is around 50 µm. Hence, at a usual supply voltage of 1–3 V, the electric field in VHalls is roughly 0.3–1 MV/m, which leads to velocity saturation, electrical nonlinearity, and inhomogeneous self-heating of the VHalls. It is known that these effects reduce the effectiveness of spinning schemes [27,28]. Therefore, single VHalls have roughly 10 times larger residual offset errors than conventional Hall plates. However, VHalls are also much smaller in chip layout than Hall plates. So, the solution is parallel connection and stacking of many VHalls [29–31]. Then, at identical chip space and current consumption, the offset of VHalls can be as low as the offset of Hall plates. It is also a widespread misconception, that VHalls have smaller SNR than Hall plates. This is only the case for shallow (3 µm) tubs with relatively large doping ($>5*10^{16}/cm^3$). For BiCMOS technologies with 5–10 µm deep epitaxial layers and doping levels below $10^{16}/cm^3$, the SNR of VHalls is comparable to that of Hall plates. Of course, new circuit architectures are needed to leverage the potential of VHalls, and there are many different proposals for VHall effect devices, so it will take a few years until the most cost-efficient variants are identified.

7.2.3 Comparison of Technologies for Magnetic Angle Sensing Devices

The advantage of AMRs over GMRs and TMRs is the absence of the pinned layer magnetization. This reduces manufacturing costs and it avoids errors originating from the reference magnetization: in GMRs and TMRs the inaccurate alignment of pinned layer magnetization and the drift of magnetization at high temperature and large applied field may lead to orthogonality errors. AMRs can be packed more densely than GMRs and TMRs, because the latter need minimum spacing when their pinned layer is magnetized to avoid cross-talk errors. AMRs can be aligned arbitrarily in many directions whereas each further direction of pinned layer magnetization in GMRs and TMRs needs an additional magnetization pulse. A drawback of AMRs over GMRs and TMRs is their smaller signal amplitude: GMRs are twice as sensitive and TMRs are 27 times more sensitive. This poses no relevant problem for modern signal conditioning circuits: for $h = 0.022$ the output of an AMR Wheatstone full bridge is 192 µV/° ($=h/2*pi/180$) at a supply voltage of 1 V. This is neither critical for pre-amp noise nor for pre-amp offset (chopped pre-amps have less than 1 µV offset error). However, the main limitation is the offset and offset drift of the AMR Wheatstone bridge itself. We assume 0.33 mV standard deviation of the bridge offset at 2 V supply: this corresponds to 0.42°. Thus, often this sensor system needs some kind of calibration. One can measure the offset (and probably also its drift over temperature) in an end-of-line test and store these calibration data in an on-chip EEPROM to correct for it—yet, this calibration is a significant cost adder and it does not account for lifetime drifts. Conversely, for continuously rotating 360° applications the system may execute an autocalibration procedure, which improves offset errors (cf. Section 7.3.1). Although GMR signals are twice as large as AMR signals, they do not sufficiently solve this problem. Finally, TMRs have such a large sensitivity that the offset-induced angle error is less than 0.2° standard deviation. Yet, for worst-case devices without calibration this may add up to 1°.

For all MRs, it is possible to sputter them on top of a silicon CMOS wafer with the sensor circuit below them. Therefore, MRs do not need additional chip space beyond the one used for the signal conditioning ASIC. Yet, since AMRs and GMRs are metallic conductors, their sheet resistance is low. Therefore, they have to be long in order to sum up to the convenient impedance levels of kiloohms, and this calls for considerable size of each resistor. As we explained previously, TMR angle sensors are similarly large. The large number of eight resistors for both Wheatstone bridges leads to sensor systems of roughly 0.5 mm in diameter. The field of small magnets is not perfectly homogeneous across such a large area. Thus, the size of the complete sensor layout generally leads to non-negligible angle errors [38].

For testing and calibration purposes, one can place conductors above/below the sensing elements and send a current through them to generate magnetic test fields. For small Hall effect devices, this is more efficient than for large MR sensors. Besides, testing Hall effect devices can be done at weak fields, because they are linear versus field strength—but for strong-field MRs, one needs to apply much larger fields, which calls for massive conductors. This effectively costs chip space because the interconnect layer is used for the test conductors and so it is not available for the chip circuit in this area, even though the MRs themselves would only float above the circuit elements.

Every now and then, it is stressed that an apparent advantage of MRs is their intrinsic angular dependence [32]: their signals vary sinusoidally with the angle of the applied field. Taking a closer look, we doubt that this argument is valid: in MR sensors the angle is computed as the arctangent of the ratio of two bridge output voltages, while in Hall sensors the angle is computed as the arctangent of the ratio of two Hall outputs—so the computational effort is the same. For the circuit, the only difference in extracting the angle is the larger signal amplitude of the MRs (see Table 7.1); however, signal amplitudes down to the microvolt (μV) range are not a blocking point for contemporary integrated circuits. The additional amount of circuit complexity for smart Hall sensors integrated with the signal conditioning circuit on the same chip need not be a relevant cost factor. MRs have a larger SNR than silicon Hall sensors, but there are only niche applications where noise is an issue. In fact, the sinusoidal dependence of MRs on the magnetic field angle is a disadvantage, because accurate information on the strength of the applied field is lost. Hall sensors directly measure the field components and so they can compute the magnitude of the applied field with an accuracy of 1% in a wide temperature range: −50°C to 150°C. Also, some MR sensors offer a rough check of the strength of the applied field: they compute the sum of both squared Wheatstone bridge outputs $\sqrt{OUT_{\cos}^2 + OUT_{\sin}^2}$, which is independent of the rotational position and proportional to temperature-dependent magnetic sensitivity h. It decreases monotonously (but not linearly) with magnetic field strength. This may be used for the detection of lost magnets, if the rotating magnet loosens from the shaft. Yet, it cannot reliably detect, if the distance between the magnet and sensor varies inappropriately on the order of 1 mm due to axial play, if ferromagnetic filings cling to the magnet and change its field, if magnet portions break off the magnet due to thermomechanical stress or collision, or if the magnet demagnetizes due to excessive mechanical shock or thermal load or corrosion. The relevant issue for the application is: Suppose the magnet changes gradually from its original strength to complete absence. First, the angle sensor can cope with a moderate decline in a magnetic field but at some point it will suffer from large error and finally it will fail. Hall sensors can detect and signal the decline in a magnetic field long before it impairs their accuracy, while MR sensors reliably detect an inappropriate magnetic field with $\sqrt{OUT_{\cos}^2 + OUT_{\sin}^2}$ only when it already causes notable angle errors. However, a system with integrated coils like in [9] can implicitly measure the field of the magnet by comparison with the known coil field, yet this costs power, chip space, and system complexity.

Historically, MR sensors were preferred for angle sensing because they were the first to accurately measure a magnetic field parallel to the chip surface, and conventional end-of-shaft angle sensors place the chip perpendicular to the rotation axis and detect the diametrical field component. We call this classical type of angle sensor a *perpendicular field angle sensor*, because it detects the magnetic field perpendicular to the rotation axis. In the late 1990s, VHall effect devices were not yet mature enough to measure the perpendicular field component with sufficient accuracy. Alternatively, one could place two chips with Hall plates perpendicular to the x- and y-directions, respectively, but this suffers from assembly tolerances and the inhomogeneous field of the magnet, which makes it less attractive than several MR-sensing elements on a single chip. Another class of sensors uses integrated magnetic concentrators (IMC) on the chip. A soft-magnetic disk of 210 µm in diameter and 24 µm thick bends the magnetic flux lines from a diametrical to a vertical direction near its perimeter, so that Hall plates can be used [33,34]. Although the flux on the Hall plates is amplified by the flux concentrator, its angle accuracy is limited by the placement tolerance and shape imperfections of the flux concentrator, its saturation and coercivity, and the inhomogeneous mechanical stress (both in- and out-of-plane) on the Hall plates near the perimeter of the flux concentrator. Besides, the Hall plates have to be fairly small to benefit from the magnetic gain of the IMC, and this increases their residual offset error. The same arrangement of Hall plates can also be used for end-of-shaft angle sensing *without* any microscopic flux concentrator. Then, the Hall plates are located on a nearly 10 times larger circle. This sensor system does not detect the diametrical field component $B_x \vec{n}_x + B_y \vec{n}_y$ but the gradients of the axial field components $\partial B_z/\partial x$, $\partial B_z/\partial y$ [35,36]. We call it the *axial field-gradient angle sensor*. The simple fact of skipping the flux concentrator completely changes the sensing principle from *perpendicular field* to *axial field gradient*: The former is severely impaired by superimposed magnetic disturbance fields while the latter cancels out homogeneous disturbances very effectively. On the other hand, perpendicular field angle sensors suffer much less from angle errors due to assembly tolerances than axial field-gradient angle sensors (see Sections 7.3 and 7.4). Meanwhile, VHall effect technology in silicon achieves SNRs comparable with silicon Hall plates and residual offsets down to 100 µT. In the next decade, VHall technology should become a serious competitor to all MR and IMC technologies, because it achieves the same accuracy without the need for individual magnetic calibration over temperature while still offering the highest versatility at the lowest costs in standard CMOS/BiCMOS technology.

Table 7.1 compares the key parameters of common magnetic angle sensor technologies. A "fair" comparison is difficult because some parameters of one technology depend on boundary conditions such as the magnet, while the same parameters of another technology do not. In such cases, we have tried to avoid extreme boundary conditions (e.g., we assume standard magnets instead of optimized ones for MRs or Hall). For strong-field MRs, signal amplitude does not depend on magnetic field strength, whereas Hall signals scale linearly. In both cases, we assume a supply voltage of 2 V. At typical fields of 45 mT, TMRs are 130 times more sensitive than silicon Hall plates without IMC. For the IMC, we assumed a realistic magnetic gain of ~5 (compared to the horizontal Hall [HHall] system *without* IMC on a 10 times larger reading radius). For VHalls, we conservatively assumed an optimized five-contact device (see also [24]). Table 7.1 assumes identical impedance for all sensors: MRs have input and output resistances of 1 kOhm for the Wheatstone bridges of the sine and the cosine channel. Hall sensors also have two channels (sin/cos) and we assume the same input and output resistances per channel. Thus, in each case, the sensing elements of both channels consume 4 mA.

Equal impedance gives equal thermal noise voltage at room temperature. Flicker noise is not quantified for MRs, although it may be significant for the tunneling effect. We denote it by "x" in Table 7.1. Flicker noise is cancelled out in Hall sensors by the spinning current scheme. For the noise in the computed angle, we assume the signals of both channels $A\cos\varphi + N_x$ and $A\sin\varphi + N_y$ with identical peak amplitude A and additive noise N_x, N_y. Per definition, it holds that $\tan(\varphi + \varphi_{noise}) = (A\sin\varphi + N_y)/(A\cos\varphi + N_x)$. We expand this equation in a Taylor series for the small angle noise, keep only the dominant linear term, and separate it at the left-hand side, $\varphi_{noise} \cong (N_y\cos\varphi - N_x\sin\varphi)/A$. If the noise of both channels is uncorrelated with equal standard deviation, one can sum up the squares of both channels and then take the square root to obtain $\text{stdev}\{\varphi_{noise}\} \cong \text{stdev}\{N_x\}/A$. It also means that angle noise is independent on the angular position. For AMRs, this leads to an additional factor 2 in the denominator $\text{stdev}\{\varphi_{noise}\} \cong \text{stdev}\{N_x\}/(2A)$. The values reported for the angle noise in Table 7.1 also include the noise of the signal conditioning circuit (pre-amp and analog-to-digital [A/D] converter), whereby we assume a conventional noise figure NF = 1 (which means that the circuit adds the same noise power as the sensing elements). In Table 7.1, we assume that the total sensor circuit behaves like a low-pass filter of first order with a corner frequency of 150 kHz, which gives an effective noise bandwidth of 150 kHz $\times (\pi/2) = 236$ kHz. We ignore the fact that AMRs need twice this bandwidth due to their $\cos(2\varphi)$ dependence. At a rotational speed of 50,000 rpm (=833 Hz), a 150 kHz first-order low-pass filter leads to a reasonably small phase-lag of only 0.32° or a delay time of 1.1 μs. This should be sufficient for control of nearly all kinds of motors—even fast ones. We see that all MRs, particularly TMRs, have much lower noise than Hall sensors due to their high signal amplitude, and this is often seen as an advantage for MRs. On the other hand, even Hall sensors have sufficiently low noise for most applications (0.035°–0.045° standard deviation without IMC).

The following three errors (offset, gain mismatch, orthogonality) are difficult to compare, because we have to distinguish between the errors "ex fab," "end of factory trim," and "with autocalibration" (cf. Section 7.3.1). Process tolerances may lead to large errors "ex fab," but in a final test at the sensor manufacturer they can be largely trimmed off in an "end of factory trim." But, of course, this requires test time, costs for precise magnetic tests, and a memory to store the trim data—and it does not cancel the errors completely; it only reduces them by a factor of ~5 due to limitations in the test procedure (position, magnetic field tolerances, limited test time, and test temperatures) and due to temperature and lifetime drift. Table 7.1 gives estimations of these errors "ex fab," because with sufficient effort they might be calibrated to comparable accuracy, where clear differences can be seen only with regard to specific application use cases. Offset errors are stochastic and so we can treat them like noise: if the errors of both channels are statistically independent and equal, the standard deviation of the angle error is equal to stdev$\{\Delta\varphi_{\text{offset}}\} \cong$ stdev$\{Off_{x,y}\}/A$ (again for AMRs, we have to replace $A \rightarrow 2A$). Table 7.1 shows that offset angle errors of TMRs are 2.7 times smaller than that of AMRs and six times smaller than that of GMRs. However, Hall sensors are four times better than TMRs. Although the IMC provides a magnetic gain of ~5, this does not improve the offset over pure HHalls due to the necessary small size of Hall plates (see Figure 2 in [33,34]). Gain mismatch between sin/cos-channels of AMRs leads to the largest errors near $\varphi = 22.5°$ with an error $d\varphi = \left(\arctan(1+\text{gain\_mismatch}) - 45°\right)/2$. For all other sensors, the worst-case angle is $\varphi = 45°$ and the error doubles. GMRs and HHalls with IMC have largest gain mismatch errors "ex fab." Orthogonality errors of HHalls with IMC, GMRs, and TMRs are significant. The large gain mismatch and orthogonality errors of IMCs come from misalignment with the HHalls. VHalls may show some orthogonality error due to the misalignment of small devices, which can be reduced by a larger number of small devices. Hysteresis errors are small for all sensors. They are sometimes corrupted by spurious iron content in the copper lead frame (Fe ~0.1%).

If offset, gain mismatch, and orthogonality are trimmed off, there is still some deviation from sinusoidal angle dependence. This is called *anisotropy error* for MR sensors and *nonlinearity* for Hall sensors. Here, TMRs tend to have a systematic dependence versus magnetic field strength, so that at a specific strength this error vanishes nominally, yet over process variations and temperature and lifetime drift some residual error still remains. Anisotropy error is mainly irregular versus rotation angle, so that it is difficult to eliminate without a detailed characterization of each individual device. This error is one reason why AMRs are still considered to be more accurate than TMRs (note that GMRs are not superior to AMRs in any parameter of Table 7.1, except that they can measure full 360° unambiguously). In principle, Hall sensors also have nonlinearity errors, yet in silicon they are negligible at practical field strengths (see also appendix B in [36]). IMCs with a diameter-thickness ratio of ~10 start saturating at diametrical fields of 60 mT [33].

7.3 Assembly Tolerances and Sensor Offset, Gain Mismatch, and Misalignment

The accuracy of magnetic angle sensors is limited by several errors (see Table 7.2). The dominant ones can be clustered into errors of the magnetic

TABLE 7.2

List of Error Sources of Magnetic Angle Sensors

Origin of Error	Description of Error
Single sensor element	Offset/zero-point error (bridge asymmetry, perming)
	Magnetic nonlinearity/anisotropy error
	Hysteresis (cw vs. ccw)
	Noise
	Finite size and shape of sensitive area
Sensor elements (sin/cos-channels or X/Y-Wheatstone bridges)	Gain mismatch
	Orthogonality error (misalignment)
	Cross-talk
	Arrangement/placement of elements in layout
Signal conditioning	Quantization error
	Algorithmic errors (e.g., zero-crossing, autocalibration)
	Time synchronization error of both channels
	Propagation delay and delay mismatch
	Finite bandwidth and bandwidth mismatch
Assembly tolerances	Chip warpage and mechanical stress
	Position and shape of microscopic flux-guide (IMC)
	Chip within sensor package (e.g., on lead frame)
	Sensor package on component board
	Sensor package against other sensor packages in through-shaft sensors
	Component board against rotation axis
	Magnet on shaft
	Shaft against rotation axis (e.g., bearing, mechanical load)
	Variation of airgap (=spacing of sensor to magnet)
Rotating magnet	Imperfect magnetization pattern
	Systematic errors of magnetic field (e.g., E-/T-shape functions)
	Partial demagnetization during lifetime (mechanical shock, too low coercivity, temp. cycling, desaccomodation)
Magnetic disturbance fields	Asynchronous fields from external disturbers
	Synchronous fields (e.g., in motor control, nonius cross-talk, dc fields)
	Nearby ferrous objects distorting field of rotating magnet

Note: Some errors, for example, offset, gain mismatch, and noise, have more than one origin. In the table they are only listed once.

field sensing elements, errors of the magnet, and mechanical assembly tolerances of the system.

- Errors of the sensing elements are mainly offset or zero-point errors and sensitivity mismatch and misalignment between sin/cos-channels or X/Y-Wheatstone bridges.
- Magnet imperfections are stochastic properties, such as misaligned direction of magnetization (e.g., if it is tilted out of the diametrical direction by inaccurate alignment of the magnetization field pulse) or inhomogeneous material composition (e.g., non-homogeneous filler content in injection-molded magnets). Slightly elliptical shapes due to systematic anisotropic sinter shrinkage or systematic non-homogeneity of the magnetic field (described by E- and T-shape functions in Equations 7.12 and 7.25) are not imperfections of a magnet—these are just aspects of suboptimal design. The following theory accounts for these aspects accurately, and one can figure out how to optimize the magnet to minimize relevant field non-homogeneities.
- Assembly tolerances are the inaccurate placement and alignment of magnet and sensing elements against the rotation axis in a statistical sense.

In the following, we focus on assembly tolerances and include errors of sensing elements, but we do not treat magnet imperfections explicitly. Although it is simple to model specific imperfections of the magnet, a general theory on all imperfections is hardly imaginable. Moreover, magnet imperfections lead to similar errors such as assembly tolerances: imagine a magnet whose left side contains less magnetic filler content than its right side—this will give similar results as a perfectly homogeneous magnet, which is slightly shifted to the right side. Also, a magnet whose magnetization is slightly tilted out of the diametrical direction will act similarly as a magnet with perfectly diametrical magnetization, which is mounted with the same small tilt on the shaft. Thus, we can take account of magnet imperfections by using rather pessimistic values for the mechanical assembly tolerances.

For perpendicular field angle sensors, the effect of magnet eccentricity or sensor eccentricity on angle error is straightforward: the magnetic field on the chip surface is inhomogeneous not only in magnitude but also in direction. Therefore, at a specific rotational position of the magnet, the angle sensor will give slightly different angle readings, if it is shifted laterally. So, we could measure the direction of the in-plane magnetic field component on the chip surface and identify a region within which the angle deviates by less than 0.5° or similar. This procedure was proposed to identify the maximum allowed eccentricity and to select good from bad magnets [37]. The following theory will show that such a method detects only a part of the problem, namely, eccentricities and not tilts (this measurement is sensitive only to the

Magnetic Angle Sensors

E function, not to the *T* function in Equations 7.22b and 7.24). It does not tell us how large the angle error will be, if the magnet or sensor chip is tilted against the rotation axis.

In fact, even a perfectly homogeneous magnetic field will lead to angle errors, if the sensor chip is tilted. Figure 7.4 shows such a case, where a homogeneous field is perpendicular to the *z*-axis and rotates around it. The tip of the field vector moves along a circle in the (B_x, B_y)-plane. However, if the chip is tilted by λ around the *y*-axis, the projection of the field on the chip surface moves along an ellipse. The difference between circle and ellipse leads to the angle error $d\varphi$ with

$$\sin d\varphi = \begin{pmatrix} \cos\varphi \\ \sin\varphi \end{pmatrix} \times \begin{pmatrix} \cos\varphi\cos\lambda \\ \sin\varphi \end{pmatrix} \frac{1}{\sqrt{\cos^2\varphi\cos^2\lambda + \sin^2\varphi}} \rightarrow d\varphi \cong \left(\lambda^2/4\right)\sin(2\varphi)$$

(7.5)

Here, we note that the error is of second order, because the tilt angle λ is squared, and this gives fairly small angle errors despite the large allowed tilts: a tilt of 5° gives only 0.11° angle error. Compared to through-shaft optical angle sensors, end-of-shaft and in-shaft magnetic angle sensors are very robust against assembly tolerances, and this is one of their main advantages for cheap mass production (cf. appendix C of [38]). Nevertheless, a number of questions arise in this respect: How do we optimize the magnets? Can we optimize the layout of the sensor elements in this respect? What is the maximum achievable angle accuracy with certain assembly tolerances? What is the probability density distribution of angle errors? Are perpendicular field angle sensors preferable over axial field-gradient angle sensors? This is the motivation for an in-depth treatment of tolerances in the following.

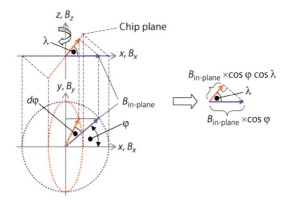

FIGURE 7.4
Angle error in a perpendicular field angle sensor caused by sensor tilt in a perfectly homogeneous diametrical magnetic field.

Here, we describe the geometrical relationships of magnet, sensor, and shaft. To this end, we need nine reference systems $\Sigma, \Sigma_1, \ldots, \Sigma_8$, which are linked by translations and rotations. We use the same conventions and symbols as in [36,38–40].

The reference frame Σ_3 for the assembly tolerances of the magnet is fixed to the shaft and centered to the rotation axis. Its rotational movement is described by the mechanical rotation angle φ around the z_3-axis:

$$\begin{pmatrix} x_4 \\ y_4 \\ z_4 \end{pmatrix} = Rz(\varphi) \begin{pmatrix} x_3 \\ y_3 \\ z_3 \end{pmatrix} \text{ with the rotation matrix } Rz(\varphi) = \begin{pmatrix} \cos\varphi & \sin\varphi & 0 \\ -\sin\varphi & \cos\varphi & 0 \\ 0 & 0 & 1 \end{pmatrix} \tag{7.6a}$$

Coordinate system Σ_4 is also centered to the rotation axis, but fixed in space. The assembly tolerances of the sensor elements are referred to Σ_4.

Turning back to the magnet, we define a coordinate system Σ that is centered with respect to the magnet and fixed to the magnet. In the absence of assembly tolerances, in this system Σ the sensor elements are ideally on the z-axis and the diametrical magnetic field on the sensor elements points in positive or negative y-direction. Eccentric mounting of the magnet on the shaft is modeled by a translation of the coordinate system $\Sigma \to \Sigma_1$:

$$\begin{pmatrix} x \\ y \\ z \end{pmatrix} = \begin{pmatrix} x_1 \\ y_1 \\ z_1 \end{pmatrix} + \begin{pmatrix} \delta_r \cos\eta \\ \delta_r \sin\eta \\ \delta_z \end{pmatrix} \tag{7.6b}$$

whereby the x-axis is parallel to the x_1-axis, the y-axis is parallel to the y_1-axis, and the z-axis is parallel to the z_1-axis. We use δ_z to model the small axial play of the shaft. Small tilts of the magnet can be around the x-axis or y-axis or around any other direction in the (x,y)-plane. We model this by first twisting the magnetization by an angle α out of its original y-direction and then tilting it by a small angle β.

$$\begin{pmatrix} x_2 \\ y_2 \\ z_2 \end{pmatrix} = Rz(\alpha) \begin{pmatrix} x_1 \\ y_1 \\ z_1 \end{pmatrix} \text{ with } Rz(\alpha) = \begin{pmatrix} \cos\alpha & \sin\alpha & 0 \\ -\sin\alpha & \cos\alpha & 0 \\ 0 & 0 & 1 \end{pmatrix} \tag{7.6c}$$

$$\begin{pmatrix} x_3 \\ y_3 \\ z_3 \end{pmatrix} = Ry(\beta) \begin{pmatrix} x_2 \\ y_2 \\ z_2 \end{pmatrix} \text{ with } Ry(\beta) = \begin{pmatrix} \cos\beta & 0 & -\sin\beta \\ 0 & 1 & 0 \\ \sin\beta & 0 & \cos\beta \end{pmatrix} \tag{7.6d}$$

$Rz(\alpha)$ and $Ry(\beta)$ are rotation matrices.

Magnetic Angle Sensors 221

Hence, from Σ_1 to Σ_4, we have three consecutive rotations that we can sum up as

$$\begin{pmatrix} x_4 \\ y_4 \\ z_4 \end{pmatrix} = A(\alpha,\beta,\varphi) \begin{pmatrix} x_1 \\ y_1 \\ z_1 \end{pmatrix} \text{ with } A(\alpha,\beta,\varphi) = Rz(\varphi)Ry(\beta)Rz(\alpha) \quad (7.6\text{e})$$

With its three parameters, $A(\alpha,\beta,\varphi)$ describes the most general rotation in 3-D space and α,β,φ are known as Euler angles [41]. Thus, we do not need any further parameters to model the most general assembly tolerances of magnets. In fact, the last rotation $Rz(\varphi)$ denotes not only the mechanical rotation but also a rotational assembly tolerance—but we do not have to separate the two, because this assembly tolerance would only mean a shift in the 0° reference position of the sensor system.

Let us compute the angle θ' between the magnetization in the magnet and its ideal plane, which is orthogonal to the rotation axis: we transfer the y-direction of the diametrical magnetization in Σ to Σ_4, which is aligned with the rotation axis:

$$A(\alpha,\beta,\varphi) \begin{pmatrix} 0 \\ 1 \\ 0 \end{pmatrix} \quad (7.6\text{f})$$

and compute the scalar product with the z_4-direction:

$$\left[A(\alpha,\beta,\varphi) \begin{pmatrix} 0 \\ 1 \\ 0 \end{pmatrix} \right] \cdot \begin{pmatrix} 0 \\ 0 \\ 1 \end{pmatrix} = \sin\beta \sin\alpha \cong \beta \sin\alpha \quad (7.6\text{g})$$

This result must be equal to $\cos(\pi/2 - \theta') = \sin\theta'$. So, we get $\theta' \cong \beta\sin\alpha$. Thus, in a Monte Carlo simulation, we may assume β to be Gaussian distributed with zero mean and then α has to be uniformly distributed in [0°,90°] or [−90°,90°] or [0°,180°] or [0°,360°].

Now, we discuss the assembly tolerances of the sensor elements. We define the coordinate system Σ_8 so that the chip surface is the (x_8, y_8)-plane, and in the absence of assembly tolerances, the rotation axis intersects the chip surface in the origin of Σ_8 with z_8 the rotation axis. Σ_8 is obtained by twisting Σ_7 by an angle ϑ around the common z-axes: $\vec{r}_8 = Rz(\vartheta)\vec{r}_7$. The angle ϑ will be of little interest to us because it just describes a misalignment of the sensor arrangement against the chip edges. Similarly to the magnet, we model the tilt of the sensor plane by two degrees of freedom: a small tilt angle λ and a large twist angle γ. We start from a reference system Σ_5 and first we twist it around z_5 by γ. This gives Σ_6, which is then tilted around y_6 to give Σ_7.

$$\begin{pmatrix} x_6 \\ y_6 \\ z_6 \end{pmatrix} = Rz(\gamma) \begin{pmatrix} x_5 \\ y_5 \\ z_5 \end{pmatrix} \text{ with } Rz(\gamma) = \begin{pmatrix} \cos\gamma & \sin\gamma & 0 \\ -\sin\gamma & \cos\gamma & 0 \\ 0 & 0 & 1 \end{pmatrix} \quad (7.7a)$$

$$\begin{pmatrix} x_7 \\ y_7 \\ z_7 \end{pmatrix} = Ry(\lambda) \begin{pmatrix} x_6 \\ y_6 \\ z_6 \end{pmatrix} \text{ with } Ry(\lambda) = \begin{pmatrix} \cos\lambda & 0 & -\sin\lambda \\ 0 & 1 & 0 \\ \sin\lambda & 0 & \cos\lambda \end{pmatrix} \quad (7.7b)$$

Again, the three Euler angles $\gamma, \lambda, \vartheta$ describe the most general rotation $A(\gamma, \lambda, \vartheta) = Rz(\vartheta) Ry(\lambda) R(\gamma)$ in 3-D space. The angle between the chip normal and the rotation axis is λ because in Σ_4 we can write

$$\left[Rz^T(\gamma) Ry^T(\lambda) Rz^T(\vartheta) \begin{pmatrix} 0 \\ 0 \\ 1 \end{pmatrix} \right] \cdot \begin{pmatrix} 0 \\ 0 \\ 1 \end{pmatrix} = \begin{pmatrix} \cos\gamma\sin\lambda \\ \sin\gamma\sin\lambda \\ \cos\lambda \end{pmatrix} \cdot \begin{pmatrix} 0 \\ 0 \\ 1 \end{pmatrix} = \cos\lambda \quad (7.7c)$$

The suffix T denotes the transpose of the rotation matrix, which is identical to its inverse. If we project the unit vector in the axial direction onto the chip surface, we get

$$Rz(\vartheta) Ry(\lambda) Rz(\gamma) \begin{pmatrix} 0 \\ 0 \\ 1 \end{pmatrix} = \begin{pmatrix} -\sin\lambda\cos\vartheta \\ \sin\lambda\sin\vartheta \\ \cos\lambda \end{pmatrix} \cong \begin{pmatrix} -\lambda\cos\vartheta \\ \lambda\sin\vartheta \\ 1 \end{pmatrix} \quad (7.7d)$$

Since the probability density of this projection should be symmetrically distributed over all four quadrants in the chip surface, ϑ must be evenly distributed between 0° and 360°. Figure 7.5 shows the meaning of the angles $\gamma, \lambda, \vartheta$: reference frame Σ_5 is aligned to the rotation axis, and in the (x_5, y_5)-plane there is the tilt axis TA. The angle γ is between the TA and the y_5-axis and the angle $\gamma + 90°$ is between the TA and the x_5-axis. Then, the rotation axis ($=z_5$-axis) is rotated around the TA by the small tilt angle λ to become the chip normal ($=z_8$-axis). Obviously, if we consider all possible tilts, the angle γ can vary between 0° and 360°, whereby a change of sign in λ is the same as if we add 180° to γ. The TA also lies in the (x_8, y_8)-plane at an angle ϑ with the y_8-axis and at an angle $90° - \vartheta$ with the x_8-axis. Here, we see that ϑ may vary between 0° and 360° for

Magnetic Angle Sensors

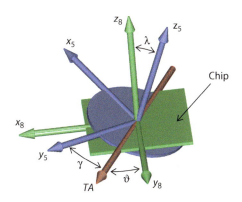

FIGURE 7.5
The three Euler angles $\gamma, \lambda, \vartheta$ for a general rotation in 3-D space. The blue reference system Σ_5 is aligned to the rotating shaft—its z_5-axis is parallel to the rotation axis. The green system Σ_8 is the chip reference system. The intersection between the (x,y)-planes of Σ_5 and Σ_8 is the red tilt axis *TA*, around which the small tilt λ is performed. The direction of the tilt axis is specified by the two angles γ, ϑ: between *TA* and both y-axes. The parameters for the drawing are $\gamma = 30°, \lambda = 20°, \vartheta = 25°$.

all possible tilts (whereby a change in the sign of λ is equal to adding 180° to ϑ).

The eccentricity of the sensor chip against the rotation axis is modeled by a translation between Σ_4 and Σ_5:

$$\begin{pmatrix} x_4 \\ y_4 \\ z_4 \end{pmatrix} = \begin{pmatrix} x_5 \\ y_5 \\ z_5 \end{pmatrix} + \begin{pmatrix} \varepsilon_r \cos\chi \\ \varepsilon_r \sin\chi \\ \varepsilon_z \end{pmatrix} \qquad (7.7e)$$

In the absence of assembly tolerances, the axial spacing between the sensor and the center of the magnet is ε_z, which is often a few millimeters (but of course there may also be a small fluctuation superimposed due to mounting tolerance).

With all these coordinate transformations, we can express the magnetic field of the permanent magnet in the chip reference frame Σ_8:

$$\vec{B}_8 = \left(\vec{B}(\vec{r}) \cdot \begin{pmatrix} \vec{n}_{x8} \\ \vec{n}_{y8} \\ \vec{n}_{z8} \end{pmatrix} \right) \cdot \begin{pmatrix} \vec{n}_{x8} \\ \vec{n}_{y8} \\ \vec{n}_{z8} \end{pmatrix} \qquad (7.7f)$$

$$\text{with } \vec{r} = \begin{pmatrix} \delta_r \cos\eta \\ \delta_r \sin\eta \\ \delta_z \end{pmatrix} + A^T(\alpha,\beta,\varphi)\left(\begin{pmatrix} \varepsilon_r \cos\chi \\ \varepsilon_r \sin\chi \\ \varepsilon_z \end{pmatrix} + A^T(\gamma,\lambda,\vartheta)\vec{r}_8 \right)$$

$$\text{and } \begin{pmatrix} \vec{n}_{x8} \\ \vec{n}_{y8} \\ \vec{n}_{z8} \end{pmatrix} = A(\gamma,\lambda,\vartheta) A(\alpha,\beta,\varphi) \begin{pmatrix} \vec{n}_x \\ \vec{n}_y \\ \vec{n}_z \end{pmatrix} \tag{7g}$$

7.3.1 Assembly Tolerances, Offset, Gain Mismatch, and Misalignment in GMRs/TMRs

Figure 7.6 shows the circuit diagram of strong-field GMR/TMR angle sensors. Four X-MRs are aligned with their pinned layer magnetizations in the x-direction and connected in order to constitute a full bridge circuit (Wheatstone bridge) and four Y-MRs are along the y-direction. Occasionally, we say sin/cos-channels instead of X/Y-bridges. If the field points in the positive x-direction, resistors R_1 and R_2 are minimum and resistors R_3 and R_4 are maximum, while resistors R_5, R_6, R_7, and R_8 are right between minimum and maximum. Consequently, the X-bridge has a negative output signal and the Y-bridge has zero output.

Yet, we have to take into account that the direction of the pinned layer magnetization is imperfectly aligned to the $\pm x$-direction for X-MRs R_1, R_2, R_3, and R_4 or to the $\pm y$-direction for Y-MRs R_5, R_6, R_7, and R_8: there is a small angular deviation τ_i for the ith resistor (Figure 7.7). The reason is that the pinned layer magnetization is defined during production by exposing the MR to a large magnetic field and a high temperature. This field may have some tiny misalignment and non-homogeneity and the hot temperature cannot be confined sharply to a single MR device without slightly affecting its

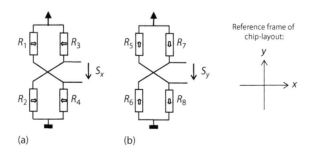

FIGURE 7.6
Schematics of the four GMRs/TMRs of the X-bridge (a) and the Y-bridge (b). The arrows in the resistors denote the direction of pinned layer magnetization in the MRs. All MRs should be close together near the point where the rotation axis intersects the chip surface.

Magnetic Angle Sensors

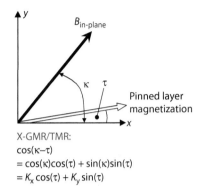
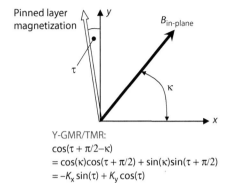

X-GMR/TMR:
cos(κ−τ)
= cos(κ)cos(τ) + sin(κ)sin(τ)
= K_x cos(τ) + K_y sin(τ)

Y-GMR/TMR:
cos(τ + π/2−κ)
= cos(κ)cos(τ + π/2) + sin(κ)sin(τ + π/2)
= −K_x sin(τ) + K_y cos(τ)

FIGURE 7.7
In a strong-field magnetic angle sensor the resistance of a GMR/TMR depends on the angle between its pinned layer magnetization and the in-plane magnetic field from the rotatable magnet.

immediate neighbors. With Equation 7.3a and Figure 7.7, we can write for the GMR resistances in Figure 7.6:

$$\text{X-bridge}: R_i = R_{\min} \frac{1+o_i}{1-h_i/2} \left\{ 1 \pm \frac{h_i}{2} \left(K_{x,i} \cos \tau_i + K_{y,i} \sin \tau_i \right) \right\} \quad (7.8a)$$

$$\text{Y-bridge}: R_i = R_{\min} \frac{1+o_i}{1-h_i/2} \left\{ 1 \pm \frac{h_i}{2} \left(K_{y,i} \cos \tau_i - K_{x,i} \sin \tau_i \right) \right\} \quad (7.8b)$$

with the plus sign for $i = 3,4,7,8$ and the minus sign for $i = 1,2,5,6$. Here, o_i denotes the offset error of the ith resistor (resistor mismatch), h_i is the magnetic sensitivity, and τ_i is the small alignment error of the pinned layer magnetization. For TMRs, we simply insert Equation 7.3b instead of Equation 7.3a into Equation 7.8a,b, which will lead to tiny differences in the results. For the sake of brevity, we write out only the GMR equations explicitly and point to the differences of TMRs only in the results. Note that $h_i \approx 0.04$ for GMRs and $h_i \approx 15 \times 0.04 = 0.6$ for TMRs. The quantities K_x and K_y in Figure 7.7 and Equation 7.8 denote the components of the normalized in-plane field:

$$K_{x,i} = \cos \kappa_i = B_x \Big/ \sqrt{B_x^2 + B_y^2} \text{ on the location of the } i\text{th MR resistor}$$

$$(i = 1, \ldots, 4) \quad (7.9a)$$

$$K_{y,i} = \sin \kappa_i = B_y \Big/ \sqrt{B_x^2 + B_y^2} \text{ on the location of the } i\text{th MR resistor}$$

$$(i = 5, \ldots, 8) \quad (7.9b)$$

Thus, in a homogeneous magnetic field, all $K_{x,i}$ are identical: $K_{x,1} = K_{x,2} = \cdots = K_{x,8}$. The same holds for $K_{y,i}$. The same holds if all MR elements are very small (point sized) and at the same location (e.g., in the center of the die). In the ideal case (i.e., without offset, gain mismatch, or orthogonality errors and in the absence of assembly tolerances and for homogeneous magnetic fields or point-sized sensor), the GMR resistances and TMR conductances are $\propto 1 \pm (h/2) K_x$ for $i = 1,2,3,4$ and $\propto 1 \pm (h/2) K_y$ for $i = 5,6,7,8$ with $K_x = B_x / \sqrt{B_x^2 + B_y^2} = \sin\varphi$ and $K_y = B_y / \sqrt{B_x^2 + B_y^2} = \cos\varphi$ and φ is the rotational position of the magnet (by convention, we assume that the magnet is aligned in the positive y-direction at $0°$ rotational position). Thus, the resistances depend on the sine and cosine of the rotation angle from which one can extract the rotational position.

However, in the general case of non-vanishing assembly tolerances and finite-sized MR sensor, $K_{x,i}$ and $K_{y,i}$ depend in a complicated way on the rotational position φ of the magnet, from the spatial distribution of the applied magnetic field, and from all assembly tolerances. In [38], we made a Taylor series expansion of $K_{x,i}$ in all small tolerance parameters $\beta, \delta_r, \lambda, \varepsilon_r$ and in the small distance r_1 of the sensor element of finite size from the rotation axis. There, we found that due to the symmetry of the system, all first-order derivatives vanish. Hence, the dominant error terms are of second order and we can immediately use the results of [38]:

$$K_{x,i} = \sigma \sin(\alpha + \gamma + \varphi + \vartheta) + k_{x,i} \text{ and } K_{y,i} = \sigma \cos(\alpha + \gamma + \varphi + \vartheta) + k_{y,i} \quad (7.10\text{a,b})$$

where σ is the sign of the B_y field at the sensor location $\sigma = B_y(0,0,\varepsilon_z + \delta_z) / |B_y(0,0,\varepsilon_z + \delta_z)| = \pm 1$. All error terms of small tolerance parameters are summed up in $k_{x,i}, k_{y,i}$:

$$k_{x,i} \cong \frac{\partial^2 K_x}{\partial \beta^2} \frac{\beta^2}{2} + \frac{\partial^2 K_x}{\partial \delta_r^2} \frac{\delta_r^2}{2} + \frac{\partial^2 K_x}{\partial \lambda^2} \frac{\lambda^2}{2} + \frac{\partial^2 K_x}{\partial \varepsilon_r^2} \frac{\varepsilon_r^2}{2} + \frac{\partial^2 K_x}{\partial r_1^2} \frac{r_1^2}{2} + \frac{\partial^2 K_x}{\partial \beta \partial \lambda} \beta\lambda$$

$$+ \frac{\partial^2 K_x}{\partial \beta \partial \varepsilon_r} \beta\varepsilon_r + \frac{\partial^2 K_x}{\partial \beta \partial \delta_r} \beta\delta_r + \frac{\partial^2 K_x}{\partial \beta \partial r_1} \beta r_1 + \frac{\partial^2 K_x}{\partial \lambda \partial \varepsilon_r} \lambda\varepsilon_r + \frac{\partial^2 K_x}{\partial \lambda \partial \delta_r} \lambda\delta_r$$

$$+ \frac{\partial^2 K_x}{\partial \lambda \partial r_1} \lambda r_1 + \frac{\partial^2 K_x}{\partial \varepsilon_r \partial \delta_r} \varepsilon_r \delta_r + \frac{\partial^2 K_x}{\partial \varepsilon_r \partial r_1} \varepsilon_r r_1 + \frac{\partial^2 K_x}{\partial \delta_r \partial r_1} \delta_r r_1 \quad (7.10\text{c})$$

For $k_{y,i}$, simply replace the index x with y in Equation 7.10c. The $k_{x,i}, k_{y,i}$ are small of second order: $|k_{x,i}| \ll 1, |k_{y,i}| \ll 1$.

$$\frac{\partial^2 K_x}{\partial \delta_r^2} = E \cos(\alpha + \gamma + \varphi + \vartheta) \sin 2\eta \quad (7.11\text{a})$$

Magnetic Angle Sensors

$$\frac{\partial^2 K_x}{\partial \varepsilon_r^2} = E\cos(\alpha+\gamma+\varphi+\vartheta)\sin 2(\alpha+\chi+\varphi) \tag{7.11b}$$

$$\frac{\partial^2 K_x}{\partial \lambda^2} = -0.5 \times \sigma \cos(\alpha+\gamma+\varphi+\vartheta)\sin 2(\alpha+\gamma+\varphi) \tag{7.11c}$$

$$\frac{\partial^2 K_x}{\partial \beta^2} = \cos(\alpha+\gamma+\varphi+\vartheta)(\sin 2\alpha)\left[\varepsilon_z(T+\varepsilon_z E)-0.5\times\sigma\right] \tag{7.11d}$$

$$\frac{\partial^2 K_x}{\partial \beta \partial \delta_r} = \cos(\alpha+\gamma+\varphi+\vartheta)\left(T\cos\alpha\sin\eta+\varepsilon_z E\sin(\alpha+\eta)\right) \tag{7.11e}$$

$$\frac{\partial^2 K_x}{\partial \beta \partial \varepsilon_r} = \cos(\alpha+\gamma+\varphi+\vartheta)\left[\cos\alpha\sin(\alpha+\chi+\varphi)T+\varepsilon_z E\sin(2\alpha+\chi+\varphi)\right] \tag{7.11f}$$

$$\frac{\partial^2 K_x}{\partial \beta \partial \lambda} = \cos(\alpha+\gamma+\varphi+\vartheta)\cos(\alpha+\chi+\varphi)(\sin\alpha)(\varepsilon_z T - \sigma) \tag{7.11g}$$

$$\frac{\partial^2 K_x}{\partial \varepsilon_r \partial \delta_r} = E\cos(\alpha+\gamma+\varphi+\vartheta)\sin(\alpha+\chi+\eta+\varphi) \tag{7.11h}$$

$$\frac{\partial^2 K_x}{\partial \lambda \partial \delta_r} = T\cos(\alpha+\gamma+\varphi+\vartheta)\cos(\alpha+\gamma+\varphi)\sin\eta \tag{7.11i}$$

$$\frac{\partial^2 K_x}{\partial \lambda \partial \varepsilon_r} = T\cos(\alpha+\gamma+\varphi+\vartheta)\cos(\alpha+\gamma+\varphi)\sin(\alpha+\chi+\varphi) \tag{7.11j}$$

Equations 7.11a through 7.11j have the factor $\cos(\alpha+\gamma+\varphi+\vartheta)$. For the Taylor series of K_y, one simply has to replace it by $(-1)\sin(\alpha+\gamma+\varphi+\vartheta)$.

In Equations 7.11a through 7.11j, the shape functions E and T describe how a non-homogeneity of the field of the permanent magnet affects the angle error. These are normalized derivatives of the magnetic field:

$$E = \frac{1}{|B_y(0,0,z)|}\frac{\partial^2 B_y(0,0,z)}{\partial x^2} \quad \text{and} \quad T = \frac{-1}{|B_y(0,0,z)|}\frac{\partial B_y(0,0,z)}{\partial z} \tag{7.12a,b}$$

at the axial position $z = \varepsilon_z + \delta_z$ of the sensor. For the ideal case of homogeneous magnetic fields, $E = T = 0$, so that many error terms of Equations 7.11a through 7.11j vanish. The T-function is more intuitive because it means that the diametrical field component of the magnet on the sensor should have zero slope versus axial distance. However, it is also highly recommended that the E-function vanishes. The meaning of the E-function is readily explained in a scan of the angle homogeneity over the chip surface according to [37]: Suppose the magnet is in position $\varphi = 0°$. Then, the in-plane component of the field on the chip (in the absence of assembly tolerances) is per definition parallel to the y-direction. If the sensor elements are shifted laterally off the rotation axis at position (x,y), they detect the angle $\tan \varphi' = B_x(x,y)/B_y(x,y)$. Thus, the angle error (or angle non-homogeneity) is $\tan \Delta\varphi = \tan(\varphi - \varphi') = \tan(0 - \varphi') = B_x(x,y)/B_y(x,y)$. We develop the field components into a Taylor series up to second order.

$$B_x(x,y) = B_x(0,0) + \frac{\partial B_x(0,0)}{\partial x} x + \frac{\partial B_x(0,0)}{\partial y} y + \frac{1}{2}\frac{\partial^2 B_x(0,0)}{\partial x^2} x^2$$

$$+ \frac{1}{2}\frac{\partial^2 B_x(0,0)}{\partial y^2} y^2 + \frac{\partial^2 B_x(0,0)}{\partial x \partial y} xy \quad (7.13a)$$

Due to the symmetry of the magnet, only the last term does not vanish [39]. Therefore, the non-homogeneity of the angle in this xy-scan* is equal to

$$\Delta\varphi \cong \frac{-1}{B_y(0,0)} \frac{\partial^2 B_x(0,0)}{\partial x \partial y} xy \to -\frac{\sigma}{2} E\varepsilon_r^2 \sin(2\chi) \quad (7.13b)$$

In Equation 7.13b, we used $\partial B_x/\partial y = \partial B_y/\partial x$, which follows from $\text{curl}\,\vec{B} = \vec{0}$, and we used $x = \varepsilon_r \cos\chi$, $y = \varepsilon_r \sin\chi$ from Equation 7.7e. This kind of (x,y)-angle scan is a method to measure the E-function and it also explains the name *eccentricity-function E* in contrast to tilt function T [39].

In general, large magnets have smaller shape functions (cf. Table III in [38]). Therefore, one should prefer large ferrite magnets over small rare-earth magnets for perpendicular field angle sensors. It is also known that magnets with stud holes facing the sensor elements can be designed to have $E = T = 0$ [40]. Such an optimization of the magnet is by far (!) the most effective way to minimize angle errors.

In Equation 7.10c, the five terms which contain r_1 account for the finite size of the sensor system. We discussed their influence on angle errors in [38]. Here, we briefly report these results:

1. The finite size of the sensor elements adds a notable error to *typical* angle errors; however, it adds only a small error to *worst-case* angle errors.

---

* It is difficult to obtain accurate values of the E-function from lateral scans of diametrical field components, because of the second derivative involved. Therefore, it is simpler to measure the angle of the diametrical field in the (x,y)-plane at a fixed position of the magnet, and fit the data to Equation 7.13b.

Magnetic Angle Sensors 229

2. The finite size of the sensor elements adds no angle error if the shape functions of the permanent magnet vanish ($E = T = 0$).
3. The layout of the sensor elements can be optimized to add no angle error due to finite size (even for magnets with non-vanishing shape functions). In such optimized layouts, the elements of each bridge must fulfill three requirements (see Figure 3 in [42], the same holds also for AMRs: see equations 21 and 22 in [40] and Figure 4 in [32]).
   a. Their gravity centers must lie on the rotation axis.
   b. Their deviation moments must vanish.
   c. Their moments of inertia around the x-axis must be equal to those around the y-axis.

In the following discussion, we assume an optimized layout so that we can skip the terms with r_1 in Equation 7.10c.

The output signals of the bridge circuits at 1 V supply voltage are

$$S_x = \frac{R_2}{R_2 + R_3} - \frac{R_4}{R_1 + R_4} \text{ and } S_y = \frac{R_6}{R_6 + R_7} - \frac{R_8}{R_5 + R_8} \quad (7.14\text{a,b})$$

We insert Equation 7.10 into Equations 7.8 and 7.14 to get the signals as functions of rotational position and tolerances. Since offsets o_i and orthogonality errors τ_i are small, we may develop the signals into a Taylor series in these parameters. For GMRs, we may consider the magnetic sensitivities h_i similarly small as assembly tolerances, yet for TMRs this is not the case. Therefore, in contrast to [38–40], we replace $h_i \to h + dh_i$ where h is the average magnetic sensitivity of all MR elements and dh_i is the sensitivity mismatch of the ith element. Then, we develop the signals into a Taylor series in dh_i. The calculation can be done in a single step, for example, with MATHEMATICA, if we multiply all small parameters $dh_1 \ldots dh_8, o_1 \ldots o_8, \tau_1 \ldots \tau_8, \beta, \lambda, \delta_r, \varepsilon_r$ with the multiplier s (s denotes "small"). Then, we develop the signals into a Taylor series in s keeping terms up to s^2. Afterward, we may take out s from the result by setting $s = 1$. This procedure gives

$$S_x = \frac{-\sigma h}{2}\sin(\alpha + \gamma + \varphi + \vartheta) + \partial_x^1 s + \partial_x^2 s^2 \text{ and}$$

$$S_y = \frac{-\sigma h}{2}\cos(\alpha + \gamma + \varphi + \vartheta) + \partial_y^1 s + \partial_y^2 s^2 \quad (7.15\text{a,b})$$

where $\partial_x^1, \partial_y^1$ are first-order and $\partial_x^2, \partial_y^2$ are second-order error terms in the tolerance parameters $dh_1 \ldots dh_8, o_1 \ldots o_8, \tau_1 \ldots \tau_8, \beta, \lambda, \delta_r, \varepsilon_r$. In the dominant terms of Equation 7.15a,b, the angles $\alpha, \gamma, \vartheta$ are added to φ. These angles are not small, so we cannot shift them into the error terms. They were used to take account of the most general tilts of magnet and sensor, because they are not only tilted around their *y*-axes, but also around any axis in their (*x,y*)-planes. So, both magnet and sensor were first twisted in their (*x,y*)-planes by α and γ, respectively, before being tilted out of their (*x,y*)-planes. ϑ is a final twist of the sensor layout in the chip plane. Of course, all these twists add up to the angle error, but they are easily calibrated out when one defines 0° at a specific reference position after the system has been assembled. This just means that we shift the zero point from $\varphi = 0°$ to $\varphi = 0° - \alpha - \gamma - \vartheta$. The exact rotational position of the shaft is the "mechanical" angle φ, but the sensor computes the angle $\varphi' = \arctan_2\{S_y, S_x\} - \alpha - \gamma - \vartheta$. We call the difference between the true angle φ and the estimated angle φ' the delta angle $\Delta\varphi$. It holds

$$\Delta\varphi \cong \tan\Delta\varphi = \tan(\varphi - \varphi') = \tan\big((\alpha + \gamma + \varphi + \vartheta) - (\alpha + \gamma + \varphi' + \vartheta)\big)$$

$$= \frac{\tan(\alpha + \gamma + \varphi + \vartheta) - \tan(\alpha + \gamma + \varphi' + \vartheta)}{1 - \tan(\alpha + \gamma + \varphi + \vartheta)\tan(\alpha + \gamma + \varphi' + \vartheta)} \quad (7.16a)$$

In Equation 7.16a, we replace $\tan(\alpha + \gamma + \varphi' + \vartheta) = \tan(S_x/S_y)$, insert Equation 7.15a,b, and make again a series expansion.

$$\Delta\varphi \cong \Delta\varphi_{const} + \Delta\varphi_{opt\,\varphi} + \sum_{n=1,2,3,4,6} \mu_n \cos\big(n(\alpha + \gamma + \varphi + \vartheta)\big) + v_n \sin\big(n(\alpha + \gamma + \varphi + \vartheta)\big)$$

$$+ \frac{\sigma}{8h}(o_1 + o_2 - o_3 - o_4)\cos(\alpha + \gamma + \varphi + \vartheta)\big(4 - h^2 \sin^2(\alpha + \gamma + \varphi + \vartheta)\big)$$

$$- \frac{\sigma}{8h}(o_5 + o_6 - o_7 - o_8)\sin(\alpha + \gamma + \varphi + \vartheta)\big(4 - h^2 \cos^2(\alpha + \gamma + \varphi + \vartheta)\big)$$

$$-\cos^2(\alpha + \gamma + \varphi + \vartheta)\left\{\frac{\tau_1 + \tau_2 + \tau_3 + \tau_4}{4} + G\sigma h \frac{\tau_1 + \tau_2 - \tau_3 - \tau_4}{8}\sin(\alpha + \gamma + \varphi + \vartheta)\right\}$$

$$-\sin^2(\alpha + \gamma + \varphi + \vartheta)\left\{\frac{\tau_5 + \tau_6 + \tau_7 + \tau_8}{4} + G\sigma h \frac{\tau_5 + \tau_6 - \tau_7 - \tau_8}{8}\cos(\alpha + \gamma + \varphi + \vartheta)\right\}$$

$$+ \frac{\sigma}{4}\cos(\alpha + \gamma + \varphi + \vartheta)\frac{2 - Gh\sin^2(\alpha + \gamma + \varphi + \vartheta)}{2 - Gh}\left(\frac{dh_1}{h} + \frac{dh_2}{h} - \frac{dh_3}{h} - \frac{dh_4}{h}\right)$$

$$- \frac{\sigma}{4}\sin(\alpha + \gamma + \varphi + \vartheta)\frac{2 - Gh\cos^2(\alpha + \gamma + \varphi + \vartheta)}{2 - Gh}\left(\frac{dh_5}{h} + \frac{dh_6}{h} - \frac{dh_7}{h} - \frac{dh_8}{h}\right)$$

$$- \frac{\sin 2(\alpha + \gamma + \varphi + \vartheta)}{8}\left(\frac{dh_1}{h} + \frac{dh_2}{h} + \frac{dh_3}{h} + \frac{dh_4}{h} - \frac{dh_5}{h} - \frac{dh_6}{h} - \frac{dh_7}{h} - \frac{dh_8}{h}\right) \quad (7.16b)$$

Magnetic Angle Sensors 231

$$\Delta\varphi_{opt\,\varphi} = \left(\lambda^2/4\right)\sin(2\alpha + 2\gamma + 2\varphi) + \lambda\beta\sin\alpha\cos(\alpha + \gamma + \varphi)$$

$$-\sigma\varepsilon_r E\bigg(\beta\varepsilon_z \sin(2\alpha + \chi + \varphi) + \delta_r \sin(\alpha + \chi + \eta + \varphi) + \frac{\varepsilon_r}{2}\sin(2\alpha + 2\chi + 2\varphi)\bigg)$$

$$-\sigma\, T\big(\lambda\cos(\alpha + \gamma + \varphi)[\beta\varepsilon_z \sin\alpha + \delta_r \sin\eta]$$

$$+\varepsilon_r \sin(\alpha + \chi + \varphi)[\beta\cos\alpha + \lambda\cos(\alpha + \gamma + \varphi)]\big) \tag{7.16c}$$

Equation 7.16b is valid for both GMRs and TMRs, whereby $G = 1$ for GMRs and $G = -1$ for TMRs. Obviously, the difference in the equation between GMRs and TMRs is small. $\Delta\varphi_{const}$ is a constant angle that does not depend on φ. $\Delta\varphi_{opt\,\varphi}$ is the angle error of GMRs and TMRs caused by assembly tolerances—it was studied in detail in [36,38–40]. The rest of Equation 7.16b gives angle errors caused merely by mismatches o_i, dh_i, τ_i of GMR/TMR elements, whereby μ_n, v_n are of second order and therefore they are small (μ_n, v_n are different for GMRs and TMRs). Lines 2–8 in Equation 7.16b give the dominant first-order GMR/TMR mismatch terms. For the standard deviation of "ex fab" angle errors due to these first-order GMR/TMR mismatch terms from Equation 7.16b, we get ~1° for GMRs and ~0.22° for TMRs with the values from Table 7.1.

$\Delta\varphi_{opt\,\varphi}$ is identical to equation 25 in [39], if we sort out the constant angle terms independent of φ in [39]. Also, Equations 7.5 and 7.13b are special cases of Equation 7.16c.

In principle, Equation 7.16b gives the angle error "ex fab" (i.e., without calibration) for every particular combination of tolerance parameters at a specific rotational angle. We are not usually interested in the angle error at a specific rotational position, but we want to judge the delta-angle in the full 0°–360° range. Typically, the plot "$\Delta\varphi$ versus φ" will be a wavy line that meanders between two extreme values $\Delta\varphi_{min}, \Delta\varphi_{max}$. Both extremes may be symmetric to zero or not—in case they are not symmetric, we can make them symmetric by changing our reference point $\varphi = 0°$. Thus, in most applications, one is mainly interested in the error band $\Delta\varphi_{max} - \Delta\varphi_{min}$. If we select the reference point $\varphi = 0°$ in an optimum way, the largest possible angle error within 360° will be $AE = \left(\Delta\varphi_{max} - \Delta\varphi_{min}\right)/2$.

7.3.1.1 Autocalibration of GMR Angle Sensors for Continuously Rotating Shafts

To make the situation even more complex, commercial sensors often use calibration routines, which are continuously executed during operation. This so-called autocalibration estimates the deviation of signals S_x, S_y from the ideal sine waves (cf. Equation 7.15a,b).

Obviously, the signals S_x, S_y suffer from offset error. It is not just depending on the o_i, because the four resistances of each bridge have different sensitivities h_i and different phase errors τ_i, so that a mixture of all these tolerances will add up to the signal offset. At a continuous rotation over 360°, the sensor circuit can estimate these offset errors, for example, by computing the signal averages:

$$o_x = \frac{1}{2\pi}\int_{\varphi=0}^{2\pi} S_x d\varphi = \left(8-h^2\right)\frac{o_1+o_2-o_3-o_4}{32} + \frac{4-h}{2-h}\frac{dh_1+dh_2-dh_3-dh_4}{16}$$

$$+\frac{dh_1^2+dh_2^2-dh_3^2-dh_4^2}{16(2-h)^2} - \left(8-h^2\right)\frac{o_1^2+o_2^2-o_3^2-o_4^2}{64}$$

$$-h\frac{(dh_1+dh_4)(o_1-o_4)+(dh_2+dh_3)(o_2-o_3)}{32}$$

$$+h\lambda\cos\vartheta\,\frac{\beta(\sigma-\varepsilon_z T)\sin\alpha-\delta_r T\sin\eta}{4}$$

$$-h\varepsilon_r\frac{\beta T\cos\alpha\sin(\chi-\gamma-\vartheta)+\beta\varepsilon_z E\sin(\alpha+\chi-\gamma-\vartheta)+\delta_r E\sin(\chi+\eta-\gamma-\vartheta)}{4}$$

(7.17)

From Equation 7.17, we obtain the average o_y of the S_y-signal by increasing the indices of o_i, dh_i by four and replacing sin → −cos and cos → −sin except for sin α, cos α, sin η. In the calculation of Equation 7.17, the sensor does not know the exact angles φ, but it can use its estimations φ′ to correct the offset error and obtain an improved estimation φ″. After several iterations, this procedure converges to the accurate value of the mechanical angle φ. Alternatively to the integration in Equation 7.17, the sensor can sample minimum and maximum values of S_x, S_y and estimate the offset by (max + min)/2. This will lead to slightly different results than Equation 7.17: in a strict sense it will not give the precise signal offset because by chance several errors may incidentally add up right near a maximum or minimum. However, in practice the differences between both offset estimation methods seem negligible, especially if the offset is small.

The signal offsets o_x, o_y in Equation 7.17 are a mixture of ohmic offsets of the MRs o_i, of magnetic sensitivity mismatches dh_i, of tilts β,λ and eccentricities ε_r, δ_r, and of non-homogeneity of the field of the permanent magnet E,T. This may be problematic if the sensor is calibrated in an end-of-line test by the semiconductor manufacturer prior to its final installation. The signal offsets o_x, o_y are not only affected by MR parameters but also by assembly tolerances and the magnet used in the end-of-line test. If they are stored in the memory of the sensor chip, they will not be exact for later use in the

Magnetic Angle Sensors

application, where a different magnet is used and different assembly tolerances occur. Therefore, it is vital that the magnet during the end-of-line test is optimized according to $\beta = E = T = 0$.

In fact, Equation 7.17 is just the first term of a Fourier series expansion of S_x, which we can write in the following form with $A_{x,n} \geq 0, A_{y,n} \geq 0$:

$$S_x = o_x + \sum_{n=1}^{\infty} A_{x,n} \sin\left(n(\alpha+\gamma+\varphi+\vartheta)+\tau_{x,n}\right) \text{ and}$$

$$S_y = o_y + \sum_{n=1}^{\infty} A_{y,n} \cos\left(n(\alpha+\gamma+\varphi+\vartheta)+\tau_{y,n}\right) \quad (7.18\text{a,b})$$

with large $A_{x,1}, A_{y,1}$, small $A_{x,n}, A_{y,n}$ for $n \geq 2$ and small o_x, o_y. Autocalibration estimates the signal amplitudes:

$$A_{x,n} = \frac{1}{\pi} \sqrt{\left(\int_0^{2\pi} S_x \cos(n\varphi) d\varphi\right)^2 + \left(\int_0^{2\pi} S_x \sin(n\varphi) d\varphi\right)^2} \quad (7.19)$$

For $A_{y,n}$, we simply replace $S_x \rightarrow S_y$ in Equation 7.19. From Equation 7.18, we get for the signal orthogonality errors:

$$\sin(\tau_{x,n} - \tau_{y,n})$$
$$= \frac{\int_0^{2\pi} S_x \sin(n\varphi) d\varphi \int_0^{2\pi} S_y \sin(n\varphi) d\varphi + \int_0^{2\pi} S_x \cos(n\varphi) d\varphi \int_0^{2\pi} S_y \cos(n\varphi) d\varphi}{\pi^2 A_{x,n} A_{y,n}}$$

(7.20a)

$$\cos(\tau_{x,n} - \tau_{y,n})$$
$$= \frac{\int_0^{2\pi} S_x \sin(n\varphi) d\varphi \int_0^{2\pi} S_y \cos(n\varphi) d\varphi - \int_0^{2\pi} S_x \cos(n\varphi) d\varphi \int_0^{2\pi} S_y \sin(n\varphi) d\varphi}{\pi^2 A_{x,n} A_{y,n}}$$

(7.20b)

The sensor constructs new orthonormal signals from Equation 7.18a,b with $\tilde{\varphi} = \varphi + \tau_{x,1}$:

$$\tilde{S}_x = \frac{S_x - o_x}{A_{x,1}} = \sin(\alpha + \gamma + \tilde{\varphi} + \vartheta) + \text{higher harmonics} \quad (7.21\text{a})$$

$$\tilde{S}_y = \frac{(S_y - o_y)/A_{y,1} - \tilde{S}_x \sin(\tau_{x,1} - \tau_{y,1})}{\cos(\tau_{x,1} - \tau_{y,1})} = \cos(\alpha + \gamma + \tilde{\varphi} + \vartheta) + \text{higher harmonics}$$

(7.21b)

where the higher harmonics are proportional to $A_{x,n}/A_{x,1}$ and $A_{y,n}/A_{y,1}$ with $n \geq 2$ according to Equation 7.18a,b. The sensor computes the angle $\tilde{\varphi}' = \arctan(\tilde{S}_x/\tilde{S}_y) - \alpha - \gamma - \vartheta$. After a lengthy calculation where we insert Equation 7.15 into Equations 7.19 through 7.21, we get the angle error $\Delta\tilde{\varphi}$ after autocalibration of the first harmonic

$$\Delta\tilde{\varphi} \cong \tan(\varphi - \tilde{\varphi}') = \Delta\tilde{\varphi}_{const} + \Delta\tilde{\varphi}_{opt\,\varphi}$$

$$+ \sum_{n=1,2,3,4,6} \tilde{\mu}_n \cos(n(\alpha + \gamma + \varphi + \vartheta)) + \tilde{\nu}_n \sin(n(\alpha + \gamma + \varphi + \vartheta))$$

$$+ \sigma h \cos(2(\alpha + \gamma + \varphi + \vartheta))$$

$$\frac{(o_1 + o_2 - o_3 - o_4)\cos(\alpha + \gamma + \varphi + \vartheta) + (o_5 + o_6 - o_7 - o_8)\sin(\alpha + \gamma + \varphi + \vartheta)}{16}$$

$$- \sigma h \sin(2(\alpha + \gamma + \varphi + \vartheta))$$

$$\frac{(\tau_1 + \tau_2 - \tau_3 - \tau_4)\cos(\alpha + \gamma + \varphi + \vartheta) + (\tau_5 + \tau_6 - \tau_7 - \tau_8)\sin(\alpha + \gamma + \varphi + \vartheta)}{16}$$

$$+ \sigma \cos(2(\alpha + \gamma + \varphi + \vartheta))$$

$$\frac{(dh_1 + dh_2 - dh_3 - dh_4)\cos(\alpha + \gamma + \varphi + \vartheta) + (dh_5 + dh_6 - dh_7 - dh_8)\sin(\alpha + \gamma + \varphi + \vartheta)}{8(2-h)}$$

(7.22a)

$$\Delta\tilde{\varphi}_{opt\,\varphi} = \Delta\varphi_{opt\,\varphi}/2$$

(7.22b)

$\Delta\tilde{\varphi}_{const}$ is a constant angle that does not depend on φ. In many cases, it is irrelevant. $\Delta\tilde{\varphi}_{opt\,\varphi}$ gives the φ-dependent angle error after autocalibration for vanishing mismatches of GMR elements $o_1 = o_2 \ldots = o_8 = dh_1 \ldots = dh_8 = \tau_1 \ldots = \tau_8 = 0$, whereas $\Delta\varphi_{opt\,\varphi}$ in Equation 7.16b is the same error *without* autocalibration. It is of second order in the small parameters of assembly tolerances (β^2, λ^2, δ_r^2, ε_r^2, $\beta\lambda, \beta\delta, \lambda\delta_r, \lambda\varepsilon h_r, \delta_r\varepsilon h_r$). Hence, *the first harmonic autocalibration of GMRs halves angle errors due to assembly tolerances*! The second line in Equation 7.22a are second-order errors caused by offset, gain mismatch, and orthogonality errors of the MRs. They are proportional to $o_i o_j, dh_i dh_j, \tau_i \tau_j, o_i dh_j, dh_i \tau_j, \tau_i o_j$ and vary versus $n\varphi$ with $n = 1,2,3,4,6$ ($\tilde{\mu}_5 = \tilde{v}_5 = 0$). Lines number three to eight in Equation 7.22a are first-order errors caused by offset, gain mismatch, and orthogonality errors of the MRs $o_1 \ldots o_8, dh_1 \ldots dh_8, \tau_1 \ldots \tau_8$. Fortunately, there are no mixed terms proportional to $o_i \beta, dh_i \lambda, \tau_i \delta_r, o_i \varepsilon_r \ldots$ in Equation 7.22a: *The angle error after autocalibration scheme is simply a linear superposition of MR mismatches and assembly tolerances.* It is a plain sum of two errors, where the first error is caused by MR mismatches in the absence of assembly tolerances, and the second error is caused by assembly tolerances in the absence of MR mismatches.

Due to its second order, the error $\Delta \tilde{\varphi}_{opt\varphi}$ is small for single error parameters (i.e., if one considers only sensor eccentricity ε_r or only magnet tilt β). However, in rare cases there are worst-case combinations of all error parameters, which are significantly larger. In Figure 8 of [38], we showed that 0.1% of all assembled systems with an optimized layout have six (!) times larger error than the median. The explanation for rare outliers are the mixed terms $\beta\lambda, \beta\delta_r, \beta\varepsilon_r, \lambda\delta_r, \lambda\varepsilon_r, \delta_r\varepsilon_r$, which make the total error larger than the sum of only four individual errors $\beta^2, \lambda^2, \delta_r^2, \varepsilon_r^2$. In practice, the combinations of tilts and eccentricities ($\beta\delta_r, \beta\varepsilon_r, \lambda\delta_r, \lambda\varepsilon_r$) give the largest angle errors, while combinations of both tilts or combinations of both eccentricities are less severe. The rare outliers are a consequence of the high degree of symmetry of this sensor arrangement: the sensor elements are located at the rotation axis, which is the symmetry center of the magnetic field. Due to this symmetry, all relevant first-order derivatives vanish and so the dominant terms in the angle error are of second order. For typical systems, this gives high accuracy, but for very rare cases the large number of second-order error terms can sum up to much larger errors.

Conversely, the first-order terms generally have a larger impact on the typical performance and they contribute less to the rare outliers. Interestingly, they are proportional to the magnetic sensitivity h. We compute the standard deviation of this linear sum of MR-mismatch terms by summing their squares and taking the square root:

$$\frac{h}{8}\sqrt{\left(\left(\frac{2\operatorname{stdev}\{dh_i/h\}}{2-h}\right)^2 + \left(\operatorname{stdev}\{o_i\}\right)^2\right)\cos^2(2\alpha + 2\gamma + 2\varphi + 2\vartheta) + \left(\operatorname{stdev}\{\tau_i\}\right)^2 \sin^2(2\alpha + 2\gamma + 2\varphi + 2\vartheta)} \quad (7.22c)$$

Equation 7.22c has a small ripple versus φ but a four times larger mean value. Inserting the values of Table 7.1 gives ~0.002° for GMRs and ~0.02° for TMRs (whereby we tacitly assume that the results for GMRs can also be applied to TMRs despite the small differences in Equation 7.3a,b). Compared to the "ex fab" values of Table 7.1, this is an improvement of factor 540 for GMRs and only 19 for TMRs (whereby we set the given worst-case values equal to six times the standard deviation).

For the second-order mismatch terms (i.e., the second line in Equation 7.22a), an evaluation is tedious due to the large number of terms. We expect rather small errors, because already the first-order terms were small.

To sum up, *autocalibration of the first harmonic very effectively suppresses offset, gain mismatch, and orthogonality errors of the MRs, but it only halves errors caused by assembly tolerances*. It should be noted that there are more elaborate schemes of autocalibration, including higher harmonics, and they lead to further improvement of angle errors. Finally, for continuous rotations, it seems only a matter of computational effort to achieve angle accuracies better than 0.1° irrespective of the magnetic field sensing technology.

7.3.2 Assembly Tolerances, Offset, Gain Mismatch, and Misalignment in AMRs, VHalls, and HHalls with IMC

Angle sensors with VHall effect devices can be treated similarly to GMRs/TMRs. VHalls also detect the in-plane magnetic field component and the system has two groups of VHalls, one that detects the x-component and one that detects the y-component of the in-plane field. These signals also suffer from offset errors (despite spinning schemes), gain mismatch, and small non-orthogonality. Again, these errors lead to first-order terms of the angle error and again autocalibration can suppress them to values far below 0.1°. Tilts and eccentricities lead to exactly (!) the same angle errors as given in Equation 7.16c.

AMR angle sensors also respond to in-plane magnetic fields. They behave similarly to GMRs/TMRs and again Equation 7.16c describes their errors caused by tilts and eccentricities [40].

Surprisingly, Equation 7.16c also applies for angle sensors with four Hall plates regularly spaced near the perimeter of a 210μm large and 24μm thick IMC: the same shape functions apply as for perpendicular field angle sensors, and so the same magnets can be used, and assembly tolerances lead to the same angle error. Yet, additional angle errors show up if the IMC gets larger or the Hall plates are not perfectly centered against the IMC. On the other hand, block-shaped magnets lead to systematic angle errors at the fourth harmonic $\sin(4\varphi)$—in this single respect the sensor resembles an *axial field-gradient angle sensor*, which needs diametrically magnetized magnets of rotational geometry (cf. Section 7.4). All these properties can be proven by an analytical calculation similar to [39] but in oblate spheroidal coordinates, where the scalar magnetic potential is developed into a series of associated Legendre polynomials evaluated for imaginary arguments.

To sum up, all perpendicular field angle sensors behave very similar with respect to assembly tolerances.

7.4 Field-Gradient Angle Sensors

Axial field-gradient angle sensors are less intuitive than perpendicular field angle sensors. The basic idea is to sample the pattern of the axial field in a plane orthogonal to the rotation axis: $B_z(x,y)$. The motivation is that the (x,y)-plane can be a chip surface and one can use ordinary Hall plates to sample the axial field. Of course, small magnets with diametrical magnetization not only generate a large diametrical field, but they also generate a useful axial field except right on the rotation axis. Thus, as long as the chip surface extends far enough from the rotation axis, a reasonable B_z field can be detected there. In fact, we need not be worried about the exact shape of the magnet—it can be very general, asymmetric, and mounted eccentrically and tilted. In any case, it will produce an asymmetric (i.e., not rotationally symmetric) B_z pattern on the chip. We may arrange a large number of Hall plates on a reading circle of 0.5–1.5 mm radius on the chip surface. The center of the reading circle should lie on the rotation axis. Obviously, the B_z field on the reading circle is a periodic function of azimuthal position ψ, and so we can express it as a Fourier series. The first harmonic of this Fourier series has a certain phase, which rotates synchronously with the shaft. This phase can be obtained via a spatial Fourier transform over $0 \leq \psi < 2\pi$. Since the sensor chip has only a finite number of N sample points, it has to approximate the Fourier series by a discrete Fourier series. Then, the first harmonic is given by

$$F_1 = \frac{1}{2\pi}\int_0^{2\pi} B_z(R=R_r,\psi)\exp(j\psi)d\psi \rightarrow \frac{1}{N}\sum_{m=0}^{N-1} B_z\left(R=R_r,\psi=\psi^{(m)}\right)\exp\left(j\psi^{(m)}\right)$$

(7.23a)

with $j=\sqrt{-1}$. The Hall plates are located on the reading radius R_r at equidistant azimuthal spacing $\psi^{(m)} = 360° \times m/N$ with $m = 0,1,\ldots, N-1$. From Equation 7.23a, the sensor computes the estimated rotation angle with $\varphi' = \arctan_2\{\text{Re}\{F_1\},\text{Im}\{F_1\}\}$. Note that this sensor principle does not require any specific properties of the magnet, except that the B_z field is sufficiently strong to be detectable. It is also obvious that a homogeneous magnetic background field superimposed on the sensor will be cancelled out in Equation 7.23a. This is a very important advantage of this type of angle sensor: it is much more robust against magnetic disturbances than the perpendicular field angle sensors discussed in earlier chapters.

However, in practice, one has to limit the number of Hall plates to $N \leq 4$: on the one hand, they consume a lot of current and use much space on the chip. On the other hand, one needs a signal path for each Hall plate because the real and imaginary parts of Equation 7.23a are linear combinations of Hall plate signals with "odd" coefficients $\cos(2\pi m/N), \sin(2\pi m/N)$. It is not economical to implement these linear combinations in an analog signal domain; instead, one should convert each Hall signal from analog to digital and compute the linear combinations Re{F_1} and Im{F_1} in the digital domain. Hence, this costs a large number of A/D converters. Therefore, one is bound to limit the number of Hall plates. However, this means that the discrete Fourier transform returns a phase that may deviate significantly from the rotation angle, depending on the B_z pattern, that is, on the magnet. An early system [35] proposes $N=4$, which gives

$$F_1 \propto B_z(R = R_r, \psi = 0°) - B_z(R = R_r, \psi = 180°)$$
$$+ j\left(B_z(R = R_r, \psi = 90°) - B_z(R = R_r, \psi = 270°)\right) \quad (7.23b)$$

The nice feature of Equation 7.23b is that we can connect opposite Hall plates so that their signals subtract and then we need only two A/D converters—one for the real part and one for the imaginary part. From Equation 7.23b it becomes obvious why we call this type of sensor system a *gradiometer*, because it responds only to gradients of the B_z field and not to absolute values of the B_z field. It cancels out homogeneous disturbance fields.

The small number of Hall plates may lead to systematic angle errors: suppose a general magnetic field pattern expressed as a Taylor series around the rotation axis $B_z(x,y) = \sum_{m=0}^{\infty} \sum_{n=0}^{\infty} c_{m,n} x^m y^n$. A rotation of the magnet is described by the rotational coordinate transformation where a point on the reading circle with coordinates $R_r \cos\psi$, $R_r \sin\psi$ transforms into the rotated coordinates $x = R_r \cos(\psi - \varphi)$, $y = R_r \sin(\psi - \varphi)$. Inserting these two equations into Equation 7.23a with $N \to \infty$ (a continuous Fourier series instead of a discrete one) shows that F_1 contains only terms proportional to $\sin\varphi$ and $\cos\varphi$, from which φ can be inferred accurately. However, if we use Equation 7.23b instead of Equation 7.23a, F_1 contains additional error terms $\sin(4\varphi), \cos(4\varphi), \sin(8\varphi), \cos(8\varphi),\ldots$, from which the dominant ones are octupoles $c_{3,0}, c_{2,1}, c_{1,2}, c_{0,3}$. In [36] it was shown by way of example that this may be relevant in practice for block-shaped magnets 6 × 6 × 3 mm (~0.7° error). In other words, *diametrically magnetized magnets need to have rotational geometry* (e.g., cylindrical, conical, spherical) to generate ideal fields of the shape of a straight plane $B_z(x,y) = c_{0,0} + c_{1,0}x + c_{0,1}y$ and this gives no systematic angle errors for axial field-gradient angle sensors with $N=4$ test points.

So, we can say that it is the principle of axial field-gradient angle sensors to detect the orientation of the B_z plane. To this end, we need at least three locations of Hall plates $N=3$, which gives

$$F_1 \propto \sqrt{3}\left(B_z(R = R_r, \psi = 120°) - B_z(R = R_r, \psi = 240°)\right)$$
$$+ j\left(2B_z(R = R_r, \psi = 0°) - B_z(R = R_r, \psi = 120°) - B_z(R = R_r, \psi = 240°)\right) \quad (7.23c)$$

Now F_1 contains systematic error terms $\sin(3\varphi), \cos(3\varphi), \sin(6\varphi), \cos(6\varphi),\ldots$, from which the dominant ones are quadrupoles $c_{2,0}, c_{1,1}, c_{0,2}$. Systems with $N=3,4,5,\ldots$ test points may also be used for through-shaft angle sensors on a much larger reading radius (cf. Figure 7.1).

We should mention that the original idea of sensing axial fields to detect the angular position of a small permanent magnet used only $N=2$ test points [43]. This was not a gradiometer and it used MAG-FETs instead of Hall plates.

Alternatively, the orientation of the B_z plane is determined by its zero crossing line $B_z(x,y) = 0$. At zero disturbance field, this line goes through the rotation axis because $c_{0,0} = 0$ for diametrically magnetized magnets. A disturbance field shifts the B_z plane up or down thereby shifting the zero-line $B_z(x,y) = 0$ parallel within the (x,y)-chip surface. The orientation of the zero-line gives the rotation angle. This system does not rely on the linearity of the sensing elements, because it only detects the zeros of B_z. One can use MAG-FETs instead of Hall plates [44]. It needs many elements densely populating a closed curve around the rotation center and so it costs a lot of chip space.

All these types of axial field-gradient angle sensors with finite N use derivatives of the magnetic field to estimate the rotation angle and this leads to larger errors caused by assembly tolerances. This is shown in detail in [36] with the same principles as in the preceding chapters. Here, we can report only the results for the condition "ex fab" (i.e., without calibration):

$$\Delta\varphi_{opt}^{axial} = \Delta\varphi_{const}^{axial} + 3(\lambda^2/4)\sin(2\alpha + 2\gamma + 2\varphi) + 3\lambda\beta\sin\alpha\cos(\alpha + \gamma + \varphi) + \lambda\beta\sin(\gamma + \varphi)$$

$$+ \frac{\varepsilon_r}{4}\tilde{E}^{axial}\left[\beta\varepsilon_z\sin(2\alpha + \chi + \varphi) + \delta_r\sin(\alpha + \chi + \eta + \varphi) + \frac{\varepsilon_r}{2}\sin(2\alpha + 2\chi + 2\varphi)\right]$$

$$+ \frac{1}{4}\tilde{T}^{axial}\left[\beta\varepsilon_r(3\sin(2\alpha + \chi + \varphi) + 2\sin(\chi + \varphi))\right.$$

$$+ \beta\varepsilon_z\lambda(3\sin(2\alpha + \gamma + \varphi) - 2\sin(\gamma + \varphi))\right]$$

$$+ \delta_r\lambda(3\sin(\alpha + \eta + \gamma + \varphi) - 2\sin(\alpha - \eta + \gamma + \varphi)) + 3\varepsilon_r\lambda\sin(2\alpha + \chi + \gamma + 2\varphi)\Big]$$

(7.24)

$\Delta\varphi_{const}^{axial}$ denotes a constant angle shift independent of φ. The shape functions in Equation 7.24 differ from the shape functions of perpendicular field angle sensors.

$$\tilde{E}^{\text{axial}} = \lim_{R \to 0} \frac{1}{B_z(R, z = \varepsilon_z)} \frac{\partial^2 B_z(R, z = \varepsilon_z)}{\partial z^2} \text{ and}$$

$$\tilde{T}^{\text{axial}} = \lim_{R \to 0} \frac{1}{B_z(R, z = \varepsilon_z)} \frac{\partial B_z(R, z = \varepsilon_z)}{\partial z} \tag{7.25}$$

Thus, axial field-gradient angle sensors prefer different magnets with $\tilde{T}^{\text{axial}} = \tilde{E}^{\text{axial}} = 0$ (see also Section 7.5) and even for these optimized magnets the angle error is larger than for perpendicular field angle sensors (note the extra factor 3 in the first line of Equation 7.24 vs. Equation 7.16c, see also [36]). Using a computer code, we searched for the maximum AE-angle error for a given set of worst-case assembly tolerances and the same magnet for both types of angle sensors: A puck magnet was 2.5mm thick with a 6mm diameter and homogeneous diametrical magnetization; the gap between sensing elements and magnet was 1.75 mm and the assembly tolerances were 5° for both tilts, 0.3 mm magnet eccentricity, and 0.5mm sensor eccentricity. Then, the worst-case "ex fab" AE-error for perpendicular field angle sensors is 2.4° and for axial field-gradient angle sensors it is 2.66°. So, the difference is only small, but this comes from the small magnet diameter, which favors axial field-gradient angle sensors. If we change the magnet diameter to 12 mm and keep all other parameters, the worst-case "ex fab" AE-errors are 0.88° and 1.74°, respectively. Here, we see that perpendicular field angle sensors can be easily designed to be much more robust against assembly tolerances than axial field-gradient angle sensors. To sum up, *in practice we have to decide whether the angle sensor is robust against disturbance fields or robust against assembly tolerances*! One possible remedy is to place both types of angle sensors on the very same chip so that they experience the same assembly tolerances [45]. Then, assembly errors can be calibrated with the perpendicular field angle sensor after installation of the system. Later, during operation in the field, the axial field-gradient angle sensor uses these calibration values.

Yet another group of *field-gradient angle sensors* uses a rotationally symmetric magnet with axial magnetization and a ferromagnetic target that is not rotationally symmetric [46]. The target may be machined into the shaft. The bias magnet can be attached to the rotating shaft or to the sensor. The rotating asymmetric field pattern is generated by the target. The sensing elements are distributed on the chip to detect the phase of the first harmonic as in Equation 7.23a, yet the field component may be axial (B_z) or diametrical (B_x, B_y). For end-of-shaft angle sensors, the latter should be even more robust against magnetic disturbances than the former.

Axial field-gradient angle sensors with $N = 4$ test points sample $\partial B_z/\partial x$ and $\partial B_z/\partial y$; however, according to Maxwell's law $\text{curl}\vec{B} = \vec{0}$ this is equivalent to $\partial B_x/\partial z$ and $\partial B_y/\partial z$. Thus, we can place a sensor chip in a leaded package upright on the rotation axis (z-axis), so that its long edge is parallel to the z-axis and its short edge is parallel to the y-axis. Two HHalls on the z-axis spaced apart by 1.5 mm sample the B_x-field and their difference gives $\partial B_x/\partial z$. Two VHalls on the z-axis spaced apart by 1.5 mm sample the B_y-field and their difference gives $\partial B_y/\partial z$. Both differences are like $\sin\varphi$ and $\cos\varphi$ from which the system can compute the rotation angle. We call this a *diametrical field-gradient angle sensor*. Obviously, it cancels out homogeneous disturbance fields.

7.5 Magnets for Magnetic Angle Sensors

In Figure 7.2, we show a classification scheme for magnetic angle sensors including common types of magnets. Basically, magnets should be cheap and supply a sufficiently strong field on the magnetic field sensing elements. But they should also provide a well-defined field pattern with sufficient homogeneity. For end-of-shaft sensors, we have to distinguish between perpendicular field angle sensors and axial field-gradient angle sensors. In the first case, magnets should apply a large diametrical field on the sensing elements located on the rotation axis, which leads to magnet diameters above 8 mm and thicknesses around 2 mm. In the second case, magnets should apply a strong axial field gradient on the sensing elements located on the 1–2 mm large reading circle around the axis of rotation, which leads to magnet diameters of less than 8 mm. In fact, the maximum B_z-field gradient is near 5 mm in diameter at ~5 mm thickness: for an axial spacing of 2 mm between sensing elements and magnet and at a remanence of 1 T the gradient is 44 mT/mm. For small shape functions E and T (cf. Equation 7.12a,b), we can simply increase the diameter of the magnets for perpendicular field angle sensors: it makes the field stronger and more homogeneous. For axial field-gradient angle sensors this does not work, because a larger magnet makes the field more homogeneous and so the gradient drops and the sensor signals drop, too! Conversely, for perpendicular field angle sensors we can even make *both* shape functions vanish, by choosing a diametrical magnetization and introducing a stud hole into this side of the magnet, which faces the sensing elements (cf. Figure 7.8a). A stud hole for axial field-gradient angle sensors was also proposed to make it robust against assembly tolerances. Yet, the solution is of no practical use, because it drastically reduces the field gradient [36]. Figure 7.8b shows an alternative version of a magnet for vanishing shape functions in axial field-gradient angle sensors. Obviously, it is a ring axially spaced apart from a concentric short cylinder. In general, optimized magnets for perpendicular field angle sensors have a simple geometry and so they can be manufactured by injection molding or by sintering, whereas for axial field-gradient angle sensors it is difficult to optimize magnets—often they can only be manufactured by injection molding, which means weaker magnets. If we want a strong diametrical field with

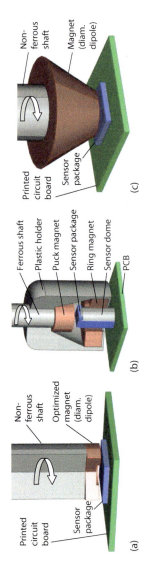

FIGURE 7.8

Various types of magnets for end-of-shaft angle sensors. (a) Diametrically magnetized puck magnet with a stud hole for perpendicular field angle sensors with $E = T = 0$ and $B_{diametral}/B_{rem} = 0.058$. The magnet is 10 mm in diameter and 2.5 mm thick; the hole is 4.72 mm in diameter and 0.5 mm in depth; the axial spacing magnet vs. XMRs/VHalls = 0.8 mm [40]. (b) Assembly of a ring and a puck magnet, both with parallel diametrical magnetization, achieves $E^{axial} = T^{axial} = 0$ at the axial field-gradient angle sensor in the center sensor dome. $B_{rem} = 0.655$ T, HcB = 496 kA/m, ring: ID = 8 mm, OD = 16 mm, 4.5 mm thick, puck: OD = 8 mm, 5 mm thick, axial distance ring vs. puck = 4.2 mm, shaft: OD = 6 mm, $\mu_r = 1500$. $B_z = 35$ mT at $R_r = 1$ mm. (c) Conical magnet with diametrical dipole magnetization achieves strong diametrical field with a comparably small magnet volume. For a strong NdFeB magnet with 1.2 T remanence, an outer diameter of 20 mm, axial thickness of 10 mm, aperture angle 60°, axial spacing magnet vs. MRs/VHalls = 2 mm, the diametrical field is ~190 mT with $E \approx -6700/m^2$, $T \approx 240/m$.

Magnetic Angle Sensors 243

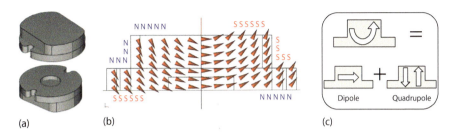

FIGURE 7.9
Magnet with diametrical dipole and axially magnetized quadrupole. (a) Top and bottom view. (b) Arc-shaped remanent magnetization and equivalent surface magnetic charges "North/South-poles" (volume charges also exist due to inhomogeneous magnetization, but they are not shown here). (c) Decomposition of arc-shaped magnetization pattern into dipole and quadrupole.

a limited amount of magnet material, the magnet should have a truncated cone geometry that can be manufactured of sintered NdFeB (cf. Figure 7.8c).

Simple block or puck magnets are often homogeneously magnetized in a diametrical direction—we call this a diametrical dipole. However, one can achieve larger fields on the sensing elements with axially magnetized quadrupoles; there, the magnetization in the right half is in the axial direction and in the left mirror symmetric half it is anti-parallel. So, they have no dipole moment* and their lowest-order multi-pole moment is a quadrupole, which has a stronger near-field and a weaker far-field. In practice, it is difficult to manufacture pure quadrupoles, so that one is bound to use a mixture of diametrical dipole and axially magnetized quadrupole; then the magnetization is arc shaped (Figure 7.9). It is preferable to use magnets with rotational symmetry as they have smaller shape functions than block magnets for equal diameter (cf. Figure 3 in [39]). Moreover, as explained in Section 7.4, axial field-gradient angle sensors with $N=4$ have a systematic fourth harmonic error for non-rotational geometries like block magnets ([36]).

The field of diametrically magnetized ring magnets is closely related to the field of short cylinders, if we think in terms of fictitious magnetic charges [47]. At the perimeter of the cylinder there are charges with a surface density equal to the scalar product of magnetization and surface normal vector: $\sigma_{\text{magnetic}} = \vec{M} \cdot \vec{n}_{\text{surface}}$. The same holds for the bore and the perimeter of diametrically magnetized rings, yet with a different sign on both surfaces, because $\vec{n}_{\text{surface}}$ points to the outside of the magnet volume. So, the ring is equal to two cylinders with anti-parallel diametrical magnetization. If the outer diameter of the ring is large, the field of the ring magnet near the axis is similar to the field of a cylinder with the ring's bore diameter, yet it points in the opposite direction. This principle has already been used in Figure 7.8b

---

* Note that the dipole moment is the volume average of the magnetization. If a magnet consists of two equal halves with anti-parallel magnetization, the volume average vanishes, and we have no dipole moment, yet a strong quadrupole moment.

to construct a field with a symmetry plane orthogonal to the rotation axis and through the magnetic field sensing elements. We can use a single ring magnet with diametrical magnetization to replace a short cylinder. The ring magnet has the advantage that its center is accessible and the slope of the field versus the axial direction vanishes by symmetry, that is, the T-function vanishes. This arrangement causes the inconvenience that the sensor has to be placed inside the ring magnet. So, the inner diameter of the ring has to be large enough to prevent collision and the sensor package has to be precisely aligned to the field, which may require a sensor dome. However, it offers the additional advantage that we may pull a cylindrical ferrous sleeve over the ring magnet. If the sleeve is an integral part of the shaft, this minimizes eccentricities and we have an in-shaft sensor. The sleeve reduces external magnetic disturbances on the sensing elements by a factor ~30 as long as the sensing elements are sufficiently deep inside the bore (at least half a bore diameter). With variation in the permeability and wall thickness of the sleeve and with two or more concentric sleeves the shielding efficiency can be increased much further [48].

This ring magnet with homogeneous diametrical magnetization also magnetizes the ferrous sleeve. The resulting magnetization pattern in the sleeve prevents flux leakage to the outside of the sleeve and it also increases the field and its homogeneity on the sensing elements (see Figure 7.10a) [49]. This brings us to the idea that the magnetization pattern of the sleeve could also be used in the magnet ring itself—then the field on the axis would increase, it would be more homogeneous, and there would no longer be a field outside the ring. Such kinds of magnetization patterns are called *Halbach magnetization* [50]. They are described by the equation:

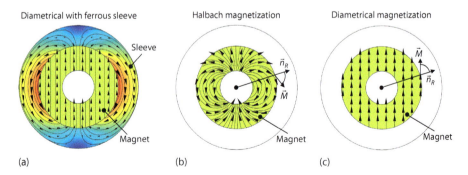

FIGURE 7.10
Magnetization vector field $\vec{M}$ in various magnet arrangements. (a) Homogeneous diametrical magnetization inside ferrous sleeve. (b) Magnet with Halbach magnetization without sleeve. (c) Magnet with homogeneous diametrical magnetization without sleeve. The magnetization in the sleeve of (a) has a similar pattern to the Halbach magnetization in (b). The angle between $\vec{n}_R$ and $\vec{M}$ (at arbitrary test points) has identical magnitude but a different sign in cases (b) and (c).

Magnetic Angle Sensors

$$\vec{M} = M_s \sin(\psi)\vec{n}_R - M_s \cos(\psi)\vec{n}_\psi = M_s \sin(2\psi)\vec{n}_x$$
$$- M_s \cos(2\psi)\vec{n}_y \ldots \text{for Halbach magnetization} \qquad (7.26a)$$

$$\vec{M} = M_s \sin(\psi)\vec{n}_R + M_s \cos(\psi)\vec{n}_\psi$$
$$= M_s \vec{n}_y \ldots \text{for homogeneous diametrical magnetization} \qquad (7.26b)$$

The difference between homogeneous diametrical magnetization in the y-direction and Halbach magnetization is simply the sign of the azimuthal component (Figure 7.10b and c). Therefore, both magnetization patterns have the same surface magnetic charges $\sigma_{\text{magnetic}} = \vec{M} \cdot \vec{n}_{\text{surface}}$, but only the Halbach magnetization has additional volume magnetic charges due to the inhomogeneous magnetization $\rho_{\text{magnetic}} = -\nabla \vec{M} = -2M_s y/(x^2 + y^2)$. Hence, the magnetic field of a Halbach magnet is given by the field of a diametrical magnet plus an additional term—this holds in air and also inside a concentric ferrous sleeve. Note that the original paper by Halbach deals with magnets with infinite axial length (pure 2-D fields). In this case, a Halbach magnet, according to Equation 7.26a, has no external field and a diametrical ring magnet, according to Equation 7.26b, has no internal field. However, if the axial length of the magnet is finite, Halbach magnets have weak external fields and diametrical ring magnets have weak internal fields. Therefore, a Halbach magnet of finite axial length does interact with the ferrous sleeve, but this interaction is small: it does not saturate the sleeve nor does it cause a notable field dependency on the poorly defined μ_r of the sleeve or on the eccentric mounting of the magnet inside the sleeve. Since the sleeve is not exposed to a strong field of the magnet, care has to be taken to avoid residual magnetization of the sleeve by other strong magnets, which would again add some small disturbance field.

The results of an experimental prototype are shown in Figure 7.11. The magnet ring has an inner diameter of 16 mm, an outer diameter of 21.8 mm, and an axial length of 16 mm. It is an anisotropic plastic injection molded ferrite material called Sprox11/21p from Magnetfabrik Bonn with $B_{\text{rem}} = 0.235$ T. It was tried in order to achieve a magnetization pattern, which is close to an internal Halbach dipole. This goal was achieved reasonably well, because the strength of the field (~57 mT) decreases by only 1.7% when the magnet is placed inside the ferrous sleeve. On the other hand, the homogeneity is good but suffers notably due to the sleeve: $E \sim 3500/\text{m}^2$ without the sleeve whereas $E \sim 5100/\text{m}^2$ inside the sleeve—this corresponds to angle errors of $AE = 0.1°$ and $AE = 0.15°$, respectively, if the sensor is placed 1 mm off the rotation axis (see Equation 7.13b). The magnet was mounted inside a sleeve with a 22 mm inner diameter and a 26 mm outer diameter made from steel DIN 1.0037-S235JR. As a sensor, we used a TLE5309D, which has two chips inside the package: one chip is mounted

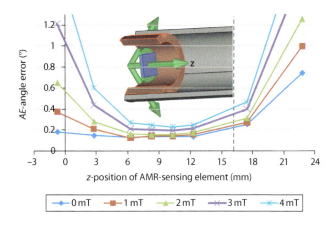

FIGURE 7.11
AE-angle error measurement of a calibrated on-axis AMR in-shaft sensor versus axial position at diametrical disturbance fields of 0, 1,..., 4 mT. Inset shows magnet flush with hollow ferrous shaft at $z = 0$ (z increases with depth inside the bore). The magnet extends from $z = 0$ up to $z = 16$ mm (denoted by the dashed vertical line). The error is 0.023° per millitesla disturbance field in the center of the magnet and it scales linearly with the disturbance field. For locations closer to the entrance or deeper in the sleeve, the angle error increases because of the non-ideal field of the magnet (see 0 mT-curve). At these locations, the magnet field weakens, so that the sensor gets more sensitive to disturbance fields that partly penetrate the sleeve.

above the die-paddle and carries a GMR angle sensor while the other chip is mounted below the die-paddle and carries an AMR angle sensor. The axial spacing of AMR and GMR is only 0.56 mm and both are on the very same radial position to avoid systematic errors caused by eccentric sensor placement. The magnet arrangement is sketched in the inset of the plot in Figure 7.11: magnet and sleeve are flush. The axial position z is zero at the entrance of the bore and increases when the test point moves inside the bore. The center of the magnet is at $z = 8$ mm. The measurement results show quite low AE-angle errors over a large axial distance $z = 6$–13 mm, whereby the AMR has smaller errors (~0.13°) than the GMR (~0.27°) when both are autocalibrated (results of GMR not shown in Figure 7.11). For sensor positions in the center of the magnet, a 4 mT diametrical disturbance field gives ~0.1° additional angle error.

References

1. Audi MediaCentre. www.audi-mediacenter.com/en/press-releases/the-innovative-shock-absorber-system-from-audi-new-technology-saves-fuel-and-enhances-comfort-6551, Aug. 10, 2016, Ingolstadt.
2. Granig, W., S. Hartmann, and B. Köppl. Performance and Technology Comparison of GMR versus Commonly Used Angle Sensor Principles for Automotive Applications. No. 2007-01-0397. SAE Technical Paper, 2007.

3. Van Der Meer, J. C., et al. A fully integrated CMOS Hall sensor with a 3.65 µT 3σ offset for compass applications. *ISSCC. 2005 IEEE International Digest of Technical Papers. Solid-State Circuits Conference, 2005*. IEEE, 2005.
4. Portelli, B., et al. Design considerations and optimization for 3-axis anisotropic magneto-resistive sensors. *Electrotechnical Conference (MELECON), 2016 18th Mediterranean*. IEEE, 2016.
5. Angleviel, D., D. Frachon, and G. Masson. Development of a Contactless Hall Effect Torque Sensor for Electric Power Steering. No. 2006-01-0939. SAE Technical Paper, 2006.
6. Koci, P. The effect of air gap, wheel speed and drive angle on the anti-lock braking system efficiency. *Carpathian Control Conference (ICCC), 2011 12th International*. IEEE, 2011.
7. Motz, M., et al. A chopped Hall sensor with small jitter and programmable "true power-on" function. *IEEE Journal of Solid-State Circuits* 40.7 (2005): 1533–1540.
8. Thomson, W. On the electro-dynamic qualities of metals: Effects of magnetization on the electric conductivity of nickel and of iron. *Proceedings of the Royal Society of London* 8 (1856): 546–550.
9. Bartos, A., A. Meisenberg, and R. Noetzel. Novel redundant magnetoresistive angle sensors. *Sensoren und Messsysteme* (2006): 99–102.
10. McGuire, T. and R. L. Potter. Anisotropic magnetoresistance in ferromagnetic 3D alloys. *IEEE Transactions on Magnetics* 11.4 (1975): 1018–1038.
11. Binasch, G., et al. Enhanced magnetoresistance in layered magnetic structures with antiferromagnetic interlayer exchange. *Physical Review B* 39.7 (1989): 4828.
12. Dieny, B., et al. Giant magnetoresistive in soft ferromagnetic multilayers. *Physical Review B* 43.1 (1991): 1297.
13. Coehoorn, R. Giant magnetoresistance in exchange-biased spin-valve layered structures and its application in read heads. In U. Hartmann (ed.) *Magnetic Multilayers and Giant Magnetoresistance*. Springer, Berlin Heidelberg, 2000. 65–127.
14. Yamazaki, H., et al. Characteristics of TMR angle sensors. *Proceedings SENSOR 2011* (2011): B8 361–365.
15. Sudo, Y., et al. Angle sensor having low waveform distortion. U.S. Patent No. 7,265,540. 4 Sep. 2007.
16. Ausserlechner, U. GMR sensors having reduced AMR effects. U.S. Patent No. 8,564,286. 22 Oct. 2013.
17. Hall, E. H. On a new action of the magnet on electric currents. *American Journal of Mathematics* 2.3 (1879): 287–292.
18. Munter, P. J. A. A low-offset spinning-current Hall plate. *Sensors and Actuators A: Physical* 22.1 (1990): 743–746.
19. Motz, M., and U. Ausserlechner. Electrical compensation of mechanical stress drift in precision analog circuits. In A. Baschirotto, P. Harpe, and K. A. A. Makinwa (eds) *Wideband Continuous-Time ΣΔ ADCs, Automotive Electronics, and Power Management*. Springer International Publishing, Cham, 2017. 297–326.
20. Ausserlechner, U. A method to compute the Hall-geometry factor at weak magnetic field in closed analytical form. *Electrical Engineering* 98.3 (2016): 189–206.
21. Ausserlechner, U. Simple formula for Hall-geometry factor of Hall-plates with 90° symmetry. *UPB Science Bulletin Series A* 78.1 (2016): 275–282.

22. Ausserlechner, U. Closed form expressions for sheet resistance and mobility from Van-der-Pauw measurement on 90° symmetric devices with four arbitrary contacts. *Solid-State Electronics* 116 (2016): 46–55.
23. Banjevic, M. High bandwidth CMOS magnetic sensors based on the miniaturized circular vertical Hall device. Thesis EPFL, Lausanne (2011).
24. Ausserlechner, U. Hall effect devices with three terminals: Their magnetic sensitivity and offset cancellation scheme. *Journal of Sensors* 2016, Article ID 5625607, 16 pages, 2016.
25. Popovic, R. S. The vertical Hall-effect device. *IEEE Electron Device Letters* 5.9 (1984): 357–358.
26. Fluitman, J. H. J. Hall-effect device with both voltage leads on one side of the conductor. *Journal of Physics E: Scientific Instruments* 13.7 (1980): 783.
27. Udo, A. Limits of offset cancellation by the principle of spinning current Hall probe. *Sensors*, 2004. *Proceedings of IEEE*. IEEE, 2004.
28. Ruther, P., et al. Thermomagnetic residual offset in integrated Hall plates. *IEEE Sensors Journal* 3.6 (2003): 693–699.
29. Banjevic, M., S. Reymond, and R. S. Popovic. On performance of series connected CMOS vertical Hall devices. Microelectronics, 2008. *MIEL 2008. 26th International Conference on*. IEEE, 2008.
30. Motz, M. and U. Ausserlechner. Vertical Hall sensor with series-connected Hall effect regions. U.S. Patent No. 9,285,439. 15 Mar. 2016.
31. Stoica, D., and M. Motz. A dual vertical Hall latch with direction detection. *ESSCIRC (ESSCIRC), 2013 Proceedings of the*. IEEE, 2013.
32. Kolbe, N., and M. Prochaska. Über den Einsatz von hochintegrierten Magnetfeldsensoren im Antriebsstrang. *GMM-Fachbericht-AmE 2010-Automotive Meets Electronics* (2010), Conference proceedings, April 15–16, 2010, VDE-Verlag, Berlin.
33. Drljaca, P. M., et al. Nonlinear effects in magnetic angular position sensor with integrated flux concentrator. *Microelectronics, 2002. MIEL 2002. 23rd International Conference on*. Vol. 1. IEEE, 2002.
34. Schott, C., R. Racz, and S. Huber. Smart CMOS sensors with integrated magnetic concentrators. *IEEE Sensors*, 2005. IEEE, 2005.
35. Metz, M., et al. Contactless angle measurement using four Hall devices on single chip. *Solid State Sensors and Actuators, 1997. TRANSDUCERS'97 Chicago., 1997 International Conference on*. Vol. 1. IEEE, 1997.
36. Ausserlechner, U. A theory of magnetic angle sensors with Hall plates and without fluxguides. *Progress in Electromagnetics Research B* 49 (2013): 77–106.
37. Blagojevic, M., N. Markovic, and R. S. Popovic. Testing the homogeneity of magnets for rotary position sensors. *IEEE Transactions on Magnetics* 50.11 (2014): 1–5.
38. Ausserlechner, U. The optimum layout for giant magneto-resistive angle sensors. *IEEE Sensors Journal* 10.10 (2010): 1571–1582.
39. Ausserlechner, U. Inaccuracies of giant magneto-resistive angle sensors due to assembly tolerances. *IEEE Transactions on Magnetics* 45.5 (2009): 2165–2174.
40. Ausserlechner, U. Inaccuracies of anisotropic magneto-resistance angle sensors due to assembly tolerances. *Progress in Electromagnetics Research B* 40 (2012): 79–99.
41. Arfken, G. *Mathematical Methods for Physicists*, 3rd ed. Academic, San Diego, CA, 1985. 198–200.

42. Schmollngruber, P., et al. GMR sensor element and its use. U.S. Patent No. 7,312,609. 25 Dec. 2007.
43. Kaulberg, T. and G. Bogason. An angle detector based on magnetic sensing. *Proceedings of IEEE International Symposium on Circuits and Systems-ISCAS'94*. Vol. 5. IEEE, 1994.
44. Kawahito, S., et al. A CMOS rotary encoder system based on magnetic pattern analysis with a resolution of 10b per rotation. *ISSCC. 2005 IEEE International Digest of Technical Papers*. Solid-State *Circuits Conference, 2005*. IEEE, 2005.
45. Ausserlechner, U. Axial and perpendicular angle sensor in single package. U.S. Patent Application No. 14/174,580.
46. Ausserlechner, U. On-axis magnetic field angle sensors, systems and methods. U.S. Patent No. 9,267,781. 23 Feb. 2016.
47. Jackson, J. D. *Classical Electrodynamics*, 2nd ed. Walter de Gruyter, Berlin, 1983. Equation (5.100).
48. Kaden, H. Wirbelströme und Schirmung in der Nachrichtentechnik. *Wirbelströme und Schirmung in der Nachrichtentechnik*, by H. Kaden. 2006 XVII, 354 S. 195 Abb. 2., vollst. neu bearb. Aufl. 1959. Nachdruck 3-540-32569-7. Springer, Berlin, 2006. Chapters C.3. and F.2.
49. Peng, Q., S. M. McMurry, and J. M. David Coey. Cylindrical permanent-magnet structures using images in an iron shield. *IEEE Transactions on Magnetics* 39.4 (2003): 1983–1989.
50. Halbach, K. Design of permanent multipole magnets with oriented rare earth cobalt material. *Nuclear Instruments and Methods* 169.1 (1980): 1–10.

Index

1/f noise, 44–45
3-D coil MIS design, 123–128
3-D microcoil technology, 108–110
Active contactless suspension (AS), 105
Agilent 4284A LCR meter, 79, 80
Ampere's law, 106
Amplitude spectrum density (ASD), 43
Anisotropic magnetoresistance (AMR), 203, 205, 209–210, 214, 234
Application-specific integrated circuits (ASICs), 207
AS, *see* Active contactless suspension (AS)
ASD, *see* Amplitude spectrum density (ASD)
ASICs, *see* Application-specific integrated circuits (ASICs)
Autonomous underwater vehicle (AUV)
    algorithms for determining FM location relative to, 153–155
    FMs monitoring from on board, 135–140
        magnetic conditions estimation, 135–137
        magnetic field analysis and estimation of distance to area, 137–139
        MS magnetometers, 139–140
        particularities, 135
    optimization algorithm for changing angular orientation, 145–151
AUV, *see* Autonomous underwater vehicle (AUV)
Axial field-gradient angle sensor, 212, 234, 238, 239
Bacterial cells, 12–13
Bacterial detection methods, 3–5
Bahreyni resonator, 68
Bi/CMOS technology, 209
BICS, *see* Built-in current sensing (BICS)
Bioanalytes detection, 57
Bio-applications, of GMR sensors, 56–58
    bioanalytes detection, 57

    biomedical signal detection, 58
    magnetic fluids monitoring, 58
Biomedical signal detection, 58
Biosensor, 56
Biosensor application, for bovine mastitis diagnosis, 1–30
    overview, 2–5
        bacterial detection methods, 3–5
        technological advances, 5
    portable "lab-on-chip" platform in raw milk, 5–17
        bacterial cells, 12–13
        biological functionalization of nanoparticles, 11–12
        calibration, 12
        cytometer platform, magnetic detection with, 8–10
        data analysis, 16–17
        mastitic milk samples, 13
        microfluidic channel fabrication, 10
        PCR reference method analysis, 13
        principles, magnetic detection, 6–8
        readout electronics, 10–11
        sterile milk samples, 13
        validation, magnetic detection, 14–16
    results, 17
        bacterial quantification evaluation, 17–18
        mAb Anti-*S. agalactiae* and pAb Anti-GB Streptococci, 18–22
        mAb Anti-*Staphylococcus* spp. and pAb Anti-*Staphylococcus aureus*, 22–24
    sensitivity and specificity, 24–30
    strengths, weaknesses, opportunities, and threats, 27–30
Bovine mastitis diagnosis, biosensor application for, 1–30
    overview, 2–5
        bacterial detection methods, 3–5
        technological advances, 5

portable "lab-on-chip" platform in raw milk, 5–17
  bacterial cells, 12–13
  biological functionalization of nanoparticles, 11–12
  calibration, 12
  cytometer platform, magnetic detection with, 8–10
  data analysis, 16–17
  mastitic milk samples, 13
  microfluidic channel fabrication, 10
  PCR reference method analysis, 13
  principles, magnetic detection, 6–8
  readout electronics, 10–11
  sterile milk samples, 13
  validation, magnetic detection, 14–16
results, 17
  bacterial quantification evaluation, 17–18
  mAb Anti-*S. agalactiae* and pAb Anti-GB Streptococci, 18–22
  mAb Anti-*Staphylococcus* spp. and pAb Anti-*Staphylococcus aureus*, 22–24
sensitivity and specificity, 24–30
strengths, weaknesses, opportunities, and threats, 27–30
Brugger magnetic field microsensor, 69–70
Built-in current sensing (BICS), 54
California Mastitis Test (CMT), 3
Capacitive detection magnetometer, 65–98
  electric model analysis, 79–81
  frequency tuning method, 85–97
    experimental validation, 95–97
    numerical resolution, 89–95
    physical modeling, 86–89
  Lorentz force–based, 67–74
    in-plane, 67–70
    out-of-plane, 70–74
  mechanical model analysis, 76–79
  overview, 66
  sensitivity evaluation, 81–85
  xylophone-based, 74–76
Circular current loops, 179–182
Circular magnet, 190–195
Clebsch–Gordan coefficients formalism, 176

CMOS, *see* Complementary metal-oxide semiconductor (CMOS)
CMT, *see* California Mastitis Test (CMT)
Co–Cu, and Co–Ag granular films, 40
Complementary metal-oxide semiconductor (CMOS), 42, 207
Comsol® Multiphysics, 82
Damping effect, 74
Data analysis, 16–17
Diamagnetic, and inductive suspension, 105, 106
Diamagnetism, 166
Diametrical field-gradient angle sensor, 239
Diametrical magnetization, 241–242
Digital compasses, 49
Dipole magnetic induction, 153
DNA detection, 57
Earnshaw–Braunbeck theorem, 105, 166
Earth's magnetic field (EMF), 136
Eddy current testing (ECT), 54–56
Electric model analysis, 79–81
Electromagnetic orientation and navigation system (EONS), 156–162
  configuration, 157–160
  orientation and positioning parameters, 160–162
Electrostatic stiffness component, 85
ELISA, *see* Enzyme-linked immunosorbent assay (ELISA)
EMF, *see* Earth's magnetic field (EMF)
Emmerich magnetometer, 67–68
End-of-shaft sensor, 200–201
Energy integrals, 174–175
Enzyme-linked immunosorbent assay (ELISA), 24, 25
EONS, *see* Electromagnetic orientation and navigation system (EONS)
Escherichia coli, 18, 22
Euler–Lagrange dynamic beam equation, 88
Fabrication, 3-D microcoils, 109–110
FEM, *see* Finite element method (FEM)
Ferromagnetic masses (FMs), 133–162
  algorithms for determining location relative to AUV, 153–155
  determining location of, 151–153

Index

electromagnetic orientation and navigation system (EONS), 156–162
  configuration, 157–160
  orientation and positioning parameters, 160–162
mobile platforms transportation, 156
monitoring from on board AUV, 135–140
  magnetic conditions estimation, 135–137
  magnetic field analysis and estimation of distance to area, 137–139
  MS magnetometers, 139–140
  particularities, 135
overview, 134–135
vector magnetic sensors calibration, 140–151
  first method, 142–143
  optimization algorithm for changing AUV angular orientation, 145–151
  second method, 144–145
Fert, A., 36
Field-gradient angle sensors, 235–239
Finite element method (FEM), 82
Flicker noise, *see* $1/f$ noise
Flow cytometers, 5
FMs, *see* Ferromagnetic masses (FMs)
Force characteristics, of magnetic suspension, 176–177
Fredholm integral equations, 169
Frequency tuning method, 85–97
  experimental validation, 95–97
  numerical resolution, 89–95
  physical modeling, 86–89
Giant magnetoresistance (GMR), and TMR sensors, 35–58, 203, 209–210, 214
  applications, 49–58
    bio-, 56–58
    general purpose magnetometers, 49
    industrial, 49–54
    non-destructive evaluation, 54–56
  assembly tolerances, offset, gain mismatch, and misalignment in, 222–234

  autocalibration for continuously rotating shafts, 229–234
  devices, 40–42
    design, 41–42
    technological issues, 40–41
  limitations, 43–48
    bandwidth, 48
    hysteresis, 46
    range of application, 43–46
    temperature drifts, 46–48
    voltage offset, 46
  overview, 36–37
  structures and phenomena, 37–40
    magnetic tunnel junctions, 39–40, 205
    other GMR structures, 40
    sandwich, 37
    spin valves, 37–39, 204
Green function, 169
Grunberg. P., 36
Halbach magnetization, 242–243
Hall effect devices, 207–209
Horizontal Hall (HHall) system, 234
Hysteresis, 46
IBD, *see* Ion beam deposition (IBD)
ICs, *see* Integrated circuits (ICs)
IMC, *see* Integrated magnetic concentrators (IMC)
IMI, *see* Intramammary infection (IMI)
Immunodiagnostics, 3
Inductive magnetization, 136
Industrial applications, of GMR sensors, 49–54
  automotive, 49
  electric current measurement, 51–54
  space, 49–51
In-plane capacitive detection, 67–70
In-shaft sensor, 200, 201
Integral equations, for secondary field, 170–171
Integrated circuits (ICs), 54
Integrated magnetic concentrators (IMC), 234
Intramammary infection (IMI), 3
Inverse dipole localization problem, 153
Ion beam deposition (IBD), 41
Johnson–Nyquist noise, *see* Thermal noise
Kadar magnetometer, 70–71

Kyynäräinen magnetometer, 72
Lagrange–Maxwell equations, 105, 178
Laplace's equation, 169, 171
Laser Doppler velocimetry (LDV), 76
Linear model, MIS, 117–119
Longitudinal, and transverse stiffness, 188, 190
Lorentz force–based capacitive magnetometers, 67–74
   in-plane detection, 67–70
   out-of-plane detection, 70–74
mAb anti-*S. agalactiae*, 18–22
mAb anti-*Staphylococcus* spp., 22–24
MAG-FETs, *see* Magnetic field-effect transistors (MAG-FETs)
Magnetic angle sensors, 199–244
   assembly tolerances, offset, gain mismatch, and misalignment, 214–234
      AMRs, VHalls, and HHalls with IMC, 234
      in GMRs/TMRs, 222–234
   field-gradient angle sensors, 235–239
   magnets for, 239–244
   overview, 200–201
   technologies, 201–214
      comparison, 209–214
      Hall effect devices, 207–209
      magnetoresistors, 202–206
Magnetic detection, 27
   with cytometer platform, 8–10
   in NDE, 54
   principles, 6–8
   validation, 14–16
Magnetic field-effect transistors (MAG-FETs), 209
Magnetic fluids, 58
Magnetic gradiometer, 153
Magnetic levitation, 165–196
   and applications, 166–168
   circular magnet, 190–195
   dynamics, 178–179
   energy integrals, 174–175
   force characteristics, 176–177
   force function of sphere in suspending field, 183–184
   integral equations for secondary field, 170–171
   one current loop, 180–182
   permanent magnets and magnetic charges, 182–183
   secondary field, 169–170
   solution for spherical shape, 171–173
   stiffness, 177–178
   superconductor in magnetic field, 168–169
   system of circular current loops, 179–180
   total field, 173–174
   two point magnets, 185–190
Magnetic permeability, 136
Magnetic sensitivity, 203, 227
Magnetic sensors (MS), for ecological monitoring, 133–162
   algorithms for determining FM location relative to AUV, 153–155
   determining location of FM, 151–153
   electromagnetic orientation and navigation system (EONS), 156–162
      configuration, 157–160
      orientation and positioning parameters, 160–162
   overview, 134–135
   stationary FMs from on board AUV, 135–140
      magnetic conditions estimation, 135–137
      magnetic field analysis and estimation of distance to area, 137–139
      MS magnetometers, 139–140
      particularities, 135
   transportation of mobile platforms to area of FM location, 156
   vector magnetic sensors calibration, 140–151
      first method, 142–143
      optimization algorithm for changing AUV angular orientation, 145–151
      second method, 144–145
Magnetic stroke, *see* Magnetic sensitivity
Magnetic suspension, *see* Magnetic levitation
Magnetic tunnel junctions (MTJ), 39–40, 41, 205

Index

Magnetoresistors (MR), 5, 202–206, 211
MALDI-TOF MS, see Matrix-assisted laser desorption ionization–time of flight mass spectrometry (MALDI-TOF MS)
MATLAB©, 10, 92
Matrix-assisted laser desorption ionization–time of flight mass spectrometry (MALDI-TOF MS), 25
Mechanical model analysis, 76–79
Meissner effect, 168
MEMS, see Microelectromechanical systems (MEMS)
Method of least squares (MLS), 145
Microelectromechanical systems (MEMS), 102–104
Microfluidic channel fabrication, 10
Micromachined electromagnetic contactless suspensions, 105–106
Micromachined Inductive Suspension (MIS), 101–128
  MEMS, 102–107
    design, MIS, 106–107
    micromachined electromagnetic contactless suspensions, 105–106
  modeling, 110–128
    analytical model, 112–119
    plane coils and 3-D coils design, 119–128
  technology, 107–110
    3-D microcoil, 108–110
    planar coils, 107–108
MLS, see Method of least squares (MLS)
Mobile platforms (MPs), 133–162
  algorithms for determining FM location relative to AUV, 153–155
  determining location of FM, 151–153
  electromagnetic orientation and navigation system (EONS), 156–162
    configuration, 157–160
    orientation and positioning parameters, 160–162
  FMs monitoring from on board AUV, 135–140
    magnetic conditions estimation, 135–137
    magnetic field analysis and estimation of distance to area, 137–139
    MS magnetometers, 139–140
    particularities, 135
  overview, 134–135
  transportation to area of FM location, 156
  vector magnetic sensors calibration, 140–151
    first method, 142–143
    optimization algorithm for changing AUV angular orientation, 145–151
    second method, 144–145
MR, see Magnetoresistors (MR)
MS, see Magnetic sensors (MS)
MS magnetometers, 139–140
MTJ, see Magnetic tunnel junctions (MTJ)
Multichannel PCB, 29
Nanoparticles (NP), 7, 27, 28
  agglomeration, 26
  biological functionalization, 11–12
National Mastitis Council protocols (NMC), 13
N-Channel MOS (NMOS), 209
NDE, see Non-destructive evaluation (NDE)
NMC, see National Mastitis Council protocols (NMC)
NMOS, see N-Channel MOS (NMOS)
Noise mechanisms, in GMR/TMR sensors
  1/f, 44–45
  random telegraph, 45
  shot, 45–46
  thermal, 43–44
Non-destructive evaluation (NDE), 54–56
  eddy current testing, 54–56
  magnetic detection, 54
Non-invasive detection system, 58
NP, see Nanoparticles (NP)
One-plane coil MIS design, 121
Out-of-plane capacitive detection, 70–74

pAb anti-GB Streptococci, 12, 17, 18–22
pAb anti-*Staphylococcus aureus*, 22–24
Panton–Valentine leukocidin, 25
Passive contactless suspension (PS), 105
Passive levitation systems, 166
PathoProof Mastitis Complete-16®, 13
PCB, *see* Printed circuit board (PCB)
PCR, *see* Polymerase chain reaction (PCR)
PDMS, *see* Polydimethylsiloxane (PDMS)
Permanent magnetization, 136
Permanent magnets, and magnetic charges, 182–183
Perpendicular field angle sensor, 212, 239
Pink noise, *see* 1/f noise
Planar coils technology, 107–108
Plane coil MIS design, 121–122
PM, *see* Proof mass (PM)
Polydimethylsiloxane (PDMS), 10
Polymerase chain reaction (PCR), 3, 16–17, 18–20, 22, 24
  reference method analysis, 13
Polytec MSA-500 laser Doppler vibrometer, 96
Popcorn noise, *see* Random telegraph noise (RTN)
Portable "lab-on-chip" platform, in raw milk, 5–17
  bacterial cells, 12–13
  biological functionalization of nanoparticles, 11–12
  calibration, 12
  cytometer platform, magnetic detection with, 8–10
  data analysis, 16–17
  mastitic milk samples, 13
  microfluidic channel fabrication, 10
  PCR reference method analysis, 13
  principles, magnetic detection, 6–8
  readout electronics, 10–11
  sterile milk samples, 13
  validation, magnetic detection, 14–16
Power spectrum density (PSD), 43
Printed circuit board (PCB), 56
Proof mass (PM), 106–107, 110, 112–23, 127
Prototheca, 22, 24
PS, *see* Passive contactless suspension (PS)

PSD, *see* Power spectrum density (PSD)
Random telegraph noise (RTN), 45
Rare earth elements (REE)-based permanent magnets, 182
Readout electronics, 10–11
REE, *see* Rare earth elements (REE)-based permanent magnets
Ren magnetic field microsensor, 73
Resonant magnetometers, 66
RTN, *see* Random telegraph noise (RTN)
Sandwich structures, in magnetic field sensing, 37
SC, *see* Superconductors (SC)
SCC, *see* Somatic cell count (SCC)
Shot noise, 45–46
Signal processing unit (SPU), 157
Silicon Hall plates, 207
Silicon-on-insulator multiuser MEMS process (SOIMUMPs), 75, 85
Soft iron distortion, 140
SOIMUMPs, *see* Silicon-on-insulator multiuser MEMS process (SOIMUMPs)
Somatic cell count (SCC), 3
Spinning current Hall probe, 208
Spin valves (SVs), 6–7, 9, 37–39, 41, 204
SPU, *see* Signal processing unit (SPU)
Stable levitation, 105, 111, 119
Staphylococcus aureus, 2, 11, 17, 22, 24
Staphylococcus epidermidis, 17
Static behavior, of MIS, 115–117
Streptococcus agalactiae, 2, 11, 12, 16–17, 18–19
Streptococcus uberis, 2, 12, 18, 19
Strong-field angle sensors, 203, 204
Superconductor, in magnetic field, 168–169
Superconductors (SC), 166
SVs, *see* Spin valves (SVs)
TFT, *see* Thin-film transistor (TFT)
Thermal noise, 43–44
Thermogravimetric method, 28
Thin-film transistor (TFT), 42
Thomson and Harsley's sensor, 68
TMR, *see* Giant magnetoresistance (GMR)
TMR structures, *see* Magnetic tunnel junctions (MTJ)

Index

Transient inductive magnetization, 137
Transient magnetic fields, 137
Tucker magnetic field microsensor, 72
Tunneling magnetoresistance (TMR) sensor, *see* Giant magnetoresistance (GMR)
Two-plane coils MIS design, 121, 122
UCL, *see* Université Catholique de Louvain (UCL)
Uniaxial wheel transport platform (UWTP), 156, 157, 158
Université Catholique de Louvain (UCL), 66
UWTP, *see* Uniaxial wheel transport platform (UWTP)
Vector magnetic sensors (VMSs) calibration, 135, 140–151
    first method, 142–143
    optimization algorithm for changing AUV angular orientation, 145–151
    second method, 144–145
    parameters estimation of background magnetic field model, 144–145
Vertical Hall effect devices (VHalls), 209, 234
VMSs, *see* Vector magnetic sensors (VMSs) calibration
Voltage offset, 46
Weak-field sensors, 204
White noise, *see* Thermal noise
Xylophone magnetometer, 66, 74–75
Young's modulus, 85–86